THE MIRROR, THE WINDOW, AND THE TELESCOPE

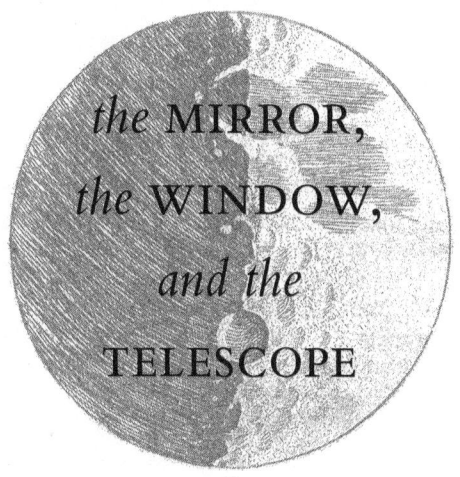

the MIRROR, the WINDOW, and the TELESCOPE

How Renaissance Linear Perspective Changed Our Vision of the Universe

SAMUEL Y. EDGERTON

Cornell University Press
Ithaca & London

Publication of this book was made possible, in part, by a grant from Williams College.

First published 2009 by Cornell University Press
First printing, Cornell Paperbacks, 2009

Printed in the United States of America

Library of Congress Cataloging-in-Publication Data

Edgerton, Samuel Y.
  The mirror, the window, and the telescope : how Renaissance linear perspec-
tive changed our vision of the universe / Samuel Y. Edgerton.
    p. cm.
  Update and sequel to: The Renaissance rediscovery of linear perspective.
  Includes bibliographical references and index.
  ISBN 978-0-8014-4758-7 (cloth : alk. paper) —
  ISBN 978-0-8014-7480-4 (pbk. : alk. paper)
    1. Perspective—History.   2. Visual perception—History.   3. Art,
Renaissance.   I. Edgerton, Samuel Y. Renaissance rediscovery of linear
perspective.   II. Title.

NC748.E34 2009
701'.8209455109024—dc22   2008031951

Cloth printing   10 9 8 7 6 5 4 3 2 1
Paperback printing   10 9 8 7 6 5 4 3

For James Ackerman

Friend in Need and Mentor in Deed

Ars sine scientia nihil est

Contents

viii   Contents

Illustrations

Preface

Illustration 1 is a sixteenth-century diagram illustrating one of the most decisive yet unappreciated ideas in the history of Western civilization: how to set out the basic geometry for drawing a picture in linear perspective.[1] In the following pages I follow up on an argument I have made elsewhere that this simple, taken-for-granted device not only changed the way Western Europeans conceived of their art during the Italian Renaissance but also had profound effects on the subsequent dramatic developments in modern science and technology. In my earlier *Renaissance Rediscovery of Linear Perspective*, first published in 1975, I tried to show how this novel concept was related to the development of accurate coordinate cartography and even to the "discovery" of America. In this volume, I end by emphasizing how it aided astronomy to discover the true form of our extraterrestrial universe.

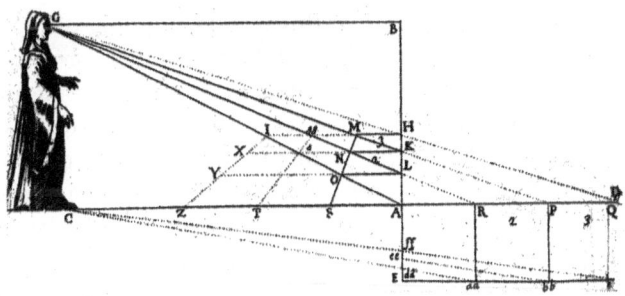

1. Perspective diagram, from Giacomo Barozzi da Vignola, *Le due regole della prospettiva pratica*, Rome, 1583.

My new book is thus both an update and a sequel. It is an update because I am adding new evidence concerning the intellectual ambience of Florence, Italy, during the early fifteenth century, in particular how the city's artisans and religious leaders grew ever more fascinated with the spiritual and moral implications of the ancient science of geometric optics.

I also submit new evidence derived from the examination of works of art from the years between 1413 and 1436 regarding the methods employed by and the date when the first Florentine artists adapted the rules of optical science to their traditional, empirical understanding of visualized nature. Included among these, just as "modern" as Masaccio and Donatello was the piously "still medieval" Fra Angelico.

My new book is also a sequel because of my concern that the subject of perspective in the arts has fallen victim to a new wave of art criticism that no longer considers it a positive idea; that it has instead actually inhibited innate artistic expression, even becoming an imperialist means to colonize the cultures of non-Western societies. Indeed, there is a tendency nowadays to downgrade the importance of perspective as merely a brief sidetrack in the evolution of world art.

Thus, I will try to reconnect the advent of perspective to its roots in the intense religious and moral preoccupations of the European late Middle Ages, to the "period eye" of the Renaissance in Michael Baxandall's famous phrase. Whatever one may say about the eventual use or misuse of geometric perspective as a tool of Western political power, it was surely conceived in the early fifteenth century as a very medieval Christian solution to a very medieval Christian problem. It must be understood in the context of the strongly held spiritual beliefs and assumptions of still devout Christians who longed for painted and sculpted images that could arouse the feeling of divine presence and reinforce their faith that God and his saints were still immanent in their daily lives.

I have found strong testimony to this in the preaching of Fra Antonino Pierozzi (1389–1459), Dominican prior of the San Marco convent (when Fra Angelico painted there) who then became the influential Archbishop of Florence. While the writings of Antonino (eventually canonized as Saint Antonine) have received some attention from modern scholars, I believe I am the first to single out his considerable views on optics, indicating just how au courant this subject was in fifteenth-century Florence. I speculate with good circumstantial reason that among the avid listeners to Antonino's popular sermons was none other than Filippo Brunelleschi, all-around artistic impresario, the first person in the history of world art to show how this science, especially the catoptrics of mirror reflection, could be accurately applied to the painting of pictures.

In the same light, I offer new analyses and interpretations of art by Masaccio, Masolino, Donatello, and Fra Angelico before and after 1425 that surely indicate that Brunelleschi must have performed his famous perspective demonstrations on or about that year. Also I will show that these artists exhibited most of the same perspective principles and shortcuts in their own works between 1425 and 1435 that Leon Battista Alberti described in his *Treatise on Painting* (*De Pictura*) after 1435, thus showing further that his own famous perspective system was basically a verbal codification of what Brunelleschi had already achieved pictorially ten years before.

Although there is much here about the persistence of religious belief, modern secularists should not be put off, because I go on to show how linear perspective itself became secularized, starting with Alberti's intention that perspective was more important to the moral rather than the religious purpose of pictures, and then almost two centuries later as applied by Galileo to his telescopic investigations of the moon's surface, eventually undermining the very medieval Christian cosmic view that had been its origin.

Finally, this book has been written with careful attention to the sensitivities of the general reader. I have tried to keep long and distracting footnotes to a minimum, and to avoid as much as possible rehearsing equally long and esoteric arguments with other scholars. My aim is to make more people, not just other specialists, more aware of just how important linear perspective was, and still is, to the uniqueness of our Western civilization.

Fortunately, I've had significant help. Indeed, I owe much to many friends and colleagues who have suggested new examples and sources to bolster my argument, and for better rhetorical ways of presenting it. Thanks most of all to Marcia Hall, Herbert Kessler, and James Elkins who read through the entire manuscript and offered most useful critical comments. I'm also indebted to Ralph Lieberman for both constructively critiquing my ideas and furnishing some much needed illustrations. Further concerning illustrations, my gratitude to Giovanni Pagliarulo, director of the Fototeca Berenson, Villa I Tatti, Florence, to Robert Volz, director, and Wayne Hammond, assistant director of the Chapin Library, Williams College, and particularly to Laurie Glover and Regina Quinn, digital scanning experts in the library of the Clark Art Institute, Williamstown, Massachusetts. Regarding the many morsels of valuable information and timely words of encouragement I often received, I must mention Marek Demianski, Zirka Filipczak, Walter Gibson, Eva Grudin, Mark Haxthausen, Frederick Ilchman, E. J. Johnson, Michael

Lewis, Peter Low, Jay Pasachoff, Linda Reynolds, Deborah Rothschild, Rocco Sinisgalli, and Stefanie Solum. My appreciation too to William Wagner, Dean of Faculty at Williams College, for granting this book a generous subvention, and to Peter Potter, Editor in Chief of Cornell University Press, for ably shepherding the manuscript through the toils of editing and printing.

Speaking of Williams College, I would also like to acknowledge the seventeen students in my last two classes, a seminar and a tutorial in which the subject was the matter of Renaissance perspective. The many discussions among myself and those named here helped mightily to shape the arguments and opinions in the following pages: Daniel Bulaevsky, David Butts, Michael Doughterty, Angelina Hong, Rachel Hooper, Charles Howard, Emilie Johnson, Tammy Kim, Alexander Mallory, Julia Nawrocki, Joshua O'Driscoll, Allison Perdue, Miranda Routh, Adam Weber, Margot Weller, Elizabeth Wilkes, and Kori Yee Litt.

Last but hardly least, let me laud my wife, Dottie, for her indispensable support and patient tolerance of my long hours of oblivious staring at the computer (*Oh la dolce prospettiva!*), and also my daughters, Perky and Mary, and son, Sam III, who likewise contributed to making this book a family project.

SAMUEL Y. EDGERTON

Williamstown, Massachusetts, April 2008

THE MIRROR, THE WINDOW, AND THE TELESCOPE

CHAPTER ONE

Introduction
Picturing the Mind's Eye

The Creator, the true first cause of geometry . . . as Plato says, always geometrizes . . . Those laws [which govern the material world] lie within the power of understanding of the human mind: God wanted us to perceive them when He created us in His image in order that we may take part in his own thoughts.

—Johannes Kepler, 1599

See the Bold-Shadow of Urania's Glory,
Immortal in His Race, no less in Story:
An Artist without Error, from whose Lyne,
Both Earth and Heav'ns, in sweet Proportion twine:
Behold Great Euclid. But, behold him well!
For 'tis in Him, DIVINITY doth dwell.

—G. Wharton, 1661

No aphorism is more taken for granted than "Seeing is Believing." All human beings, both male and female of whatever race or ethnic heritage, have the same mechanically structured eyes, and, as far as the pure physics of vision are concerned, "see" identically. Even if individual vision happens to dysfunction for some physiological reason, it can usually be corrected by spectacles that work predictably and uniformly whatever the wearer's culture, thanks to the universal scientific laws of optical geometry first devised by the ancient Greek philosopher, Euclid (fl. 300 BC). Unlike so many ideas

1

of the other great classical thinkers, Euclid's laws have never been altered or challenged, at least for explaining how we humans measure Mother Earth with our eyes. In other words, the healthy eyes of every normal person in the world perceive the same raw forms of images in the same perspective distance and size relationships just as the ancients demonstrated—that is, before the nurtured brain makes any psychological or ideological judgment as to their meaning.

Furthermore, every human being from Pleistocene times to the present has experienced in vision the apparent convergence of parallel edges of objects as they extend away from our eyes and seem to come together in a single "vanishing point" on the distant horizon, as in the classic example of photographed railroad tracks in illustration 2.

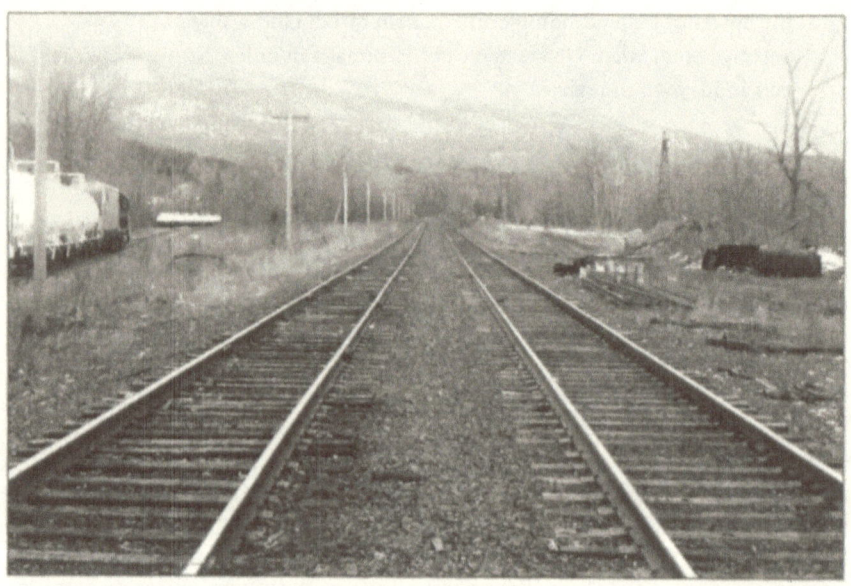

2. Railroad tracks. Photo: author's collection.

How curious it is then that, even though we see "vanishing point" perspective in phenomenal nature, the notion of rendering it in pictures is not inherent. Human beings generally, even those born with so-called artistic talent, are never automatically inclined to draw images in the same way we perceive them perspectivally—that is, according to the geometric-optical laws of vision.[1] Geometric-optical linear perspective drawing is a special skill that must be *learned*, just like reading and writing in school.[2] No matter how obvious the optical illusion of perspective convergence and its generally taken-for-granted assumption in the Western-influenced world as being the trademark of pictorial "realism," it has rarely, and almost never outside this Western-influenced world, been of concern to artists before the Italian Renaissance. Most non-Western artists, and even Western artists during the early Middle Ages, tended to render objects in "divergent perspective." Tables, for example, were often depicted wider at the more distant end than at the nearer, simply because the artist preferred such a tilted area in order to display more objects, each of which should be viewed according to its own most characteristic aspect; dinner plates, for instance, as if seen from directly above, and pitchers and glasses as if observed from the side (ill. 3).

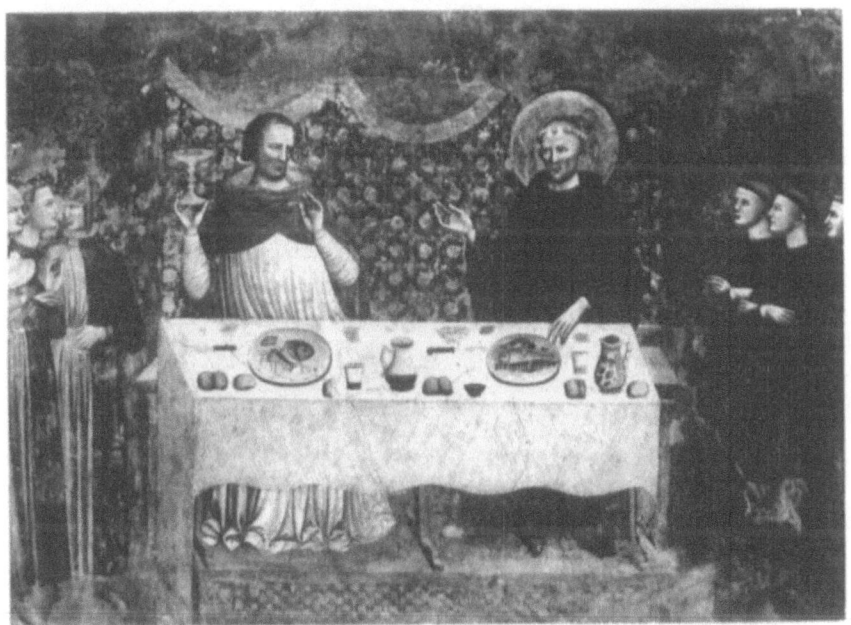

3. Anonymous Rimini painter, *Miracle of St. Guido*, Pomposa Abbey, Ferrara, ca. 1316. Fototeca Berenson, Villa I Tatti, Florence.

Such a "naïve" mind's-eye view seems indeed to be instinctive in all human beings. Our primal artistic imagination is also predisposed to separate that which is of most importance to the picture's message from that which is only incidental. This is usually accomplished by making the important figures larger than the others, with forms of objects supposed to be nearer at the bottom, and further away at the top, but with no particular size or scale differentiation due to separation by actual distance.[3]

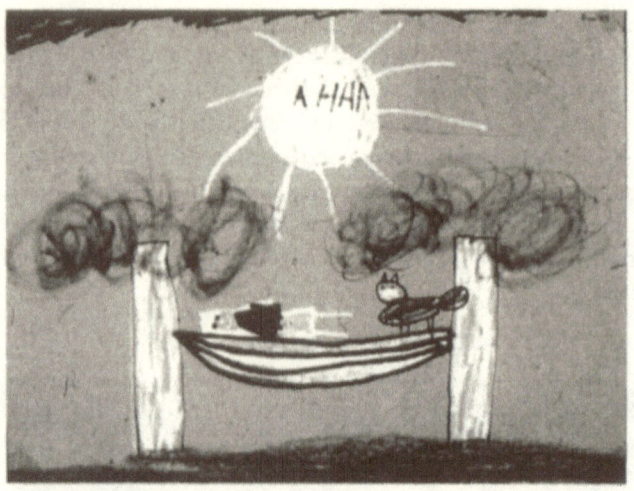

4. Drawing. Anna Filipczek, age five. Author's collection.

All this is very evident in the art of children when they first begin to make pictures. Here (ill. 4) is a drawing by a five year old that clearly shows these same natural tendencies. The subject is the young artist's mother sleeping in a hammock with her pet cat perched at one end. Note how the child has drawn the mother as if suspended above the hammock, not nestled within it. The five year old has instinctively realized that in order for the mother to be visible in the picture, she cannot be occluded by the hammock, even though that was not the way she would naturally see her mother unless she was up in the tree looking down.

It is exactly this "naïve" manner of representing visual "reality" that remains the basis of even the most sophisticated adult art everywhere in the world not influenced by Western perspective. Yet, even though Western art itself was once just as "naïve," many modern Westerners still tend to judge even the most sophisticated adult art of non-Western societies unfamiliar with Renaissance-style "realism" as "childlike," unconsciously if nonetheless pejoratively synonymous with "primitive." Many Westerners are too certain that because perspective is so rooted in scientific geometry, the "realism" it produces must be universally absolute. Its very "invention" in the West is often and unfortunately taken for granted as but another example of Western scientific superiority.

Ironically, the advent of geometric-optical perspective in art, which occurred during the still medieval and still devout Italian Renaissance, was thoroughly inspired by the quite unmodern spiritual assumptions of European Christians. However, it should also be understood that even though the ability to draw linear perspective is not instinctive, the neural connections in all human brains do innately sense geometric patterns in natural shapes, and do stimulate the desire to copy these artistically. Although this propensity is usually manifested as abstract decoration in the arts of almost every non-Western civilization, it also is at the root of Western art at the time of the Renaissance. In truth, the optical inception of geometric perspective was always just as much a neural response as any of the visual forms that led to the pictorial conventions conceived by other civilizations. Only under the peculiar demands of Western classical and Christian culture did it become "convergent perspective" and thus the unique "symbolic form" of the Renaissance, as Erwin Panofsky described it many years ago.[4]

Some time early in the year 1425 the Florentine sculptor, engineer, architect, and all-around artisan-impresario Filippo Brunelleschi (1377–1446) painted two pictures, one smaller (about eleven-and-half inches square) showing the eastern façade of the Florentine Baptistery viewed frontally from the western portal of the city's Cathedral, and the other a larger panel of Florence's nearby government headquarters known as the Palazzo Signoria, viewed obliquely from a northwest direction in the city's main public piazza.

Unfortunately, both paintings have been lost since the mid-fifteenth century. Their original subjects, however, are known to us through later verbal descriptions, but only one of which was by a witness who claimed to have actually seen the pictures. Nevertheless, based on this and the other secondhand accounts, several reconstructions have been proposed by modern scholars, including my own. These two lost panels, so far studied by only a few specialized scholars and hardly attracting the interest of even the most passionate lovers of Italian painting, were surely the most influential artworks produced during the entire European Renaissance.

How dare I make such a claim? Because the geometric linear perspective scheme employed for the first time ever in each of these modest pictures utterly changed how artists of Western civilization represented "reality" for the next four hundred years and more. Even when artists consciously violated perspective rules, they were acknowledging its cultural importance. And even more astounding is the fact that the geometric-optical perspective method Brunelleschi first introduced here has managed also to change the very way people, especially those in Westernized societies, visually verify the phenomenal world in their acculturated mind's eye. In other words, Brunelleschi's perspective not only altered how we represent what we see but how we actually see a priori.

Within weeks after Brunelleschi's initial demonstration, his perspective method was adopted by some of the most talented artists of Florence—Masaccio, Donatello, and even the conservative Fra Angelico, thence throughout Italy, and by 1600 it was being learned and adapted by artists everywhere in transalpine Western Europe. By 1700, the perspective way of perceiving visual "reality" in the physical world was accepted as a universal, natural truth, as absolute as Isaac Newton's recently proven law of gravity. It is difficult for modern eyes, long since overexposed to every possible perspective trompe l'oeil sensation from 3-D Cineramas to Disneyland holographic ghosts, to be thrilled again by Renaissance-style demonstrations of perspective "realism." Remember too, that pictorial perspective, unlike aural music (which is always just as loud no matter how often it's repeated), becomes quickly déjà vu with too frequent repetition (how many times can the eye be fooled by the same fake doorway painted open on the wall?). We have quite forgotten the startled excitement of those first Florentine viewers who had never seen a precise geometric-optical reproduction in a painted picture of what the fixed eye sees phenomenally. In a later chapter, we will examine certain paired perspective scenes, each pair created by a different artist, but with one from each pair dated shortly before Brunelleschi's demonstrations in 1425, and the second shortly after. The obvious difference between the two is what I will call the "excitement gauge." The earlier works are characterized by indifferent, even clumsy attempts at "perspective" illusionism, but then a sudden tour de force fascination, even obsession, takes over in the later examples.

Curiously, however, the original public test as to how well Brunelleschi's first perspective image resembled the real Baptistery, according to the most detailed fifteenth-century account, was by comparing the picture not to the building itself, but to the building's reflection in a mirror. As the title of this book hints, Brunelleschi's mirror will be revealed as the key component signifying why he conceived of geometric-optical linear perspective in the first place, especially in the still profoundly religious ambience of medieval

Florence. Brunelleschi employed the mirror for the same reason as expressed by Saint Paul in his famous First Epistle to the Corinthians 13:12, "At present we see indistinctly, as in a mirror, but then [in heaven] face to face."

Indeed, Brunelleschi's remarkable achievement, as he first understood it, was not the opening of the door to secular observation of objective nature as has customarily been claimed, but ironically a last gasp of the spiritual Middle Ages. It has been said that the Renaissance was the most medieval thing the Middle Ages invented. Indeed, the fifteenth century was a time still yearning for a new means to stimulate the Faith after a bitter two centuries of political turmoil in the Church (the Schism) and ideological frustration (the failure of the Crusades). There was a desperate need to reinspire Western Christians to stem expanding Islam, and even recapture Jerusalem. Linear perspective should therefore retool the visual arts, refurbishing them to present the Christian message more convincingly and help shore up the sagging beliefs of an increasingly cynical population.

The second cofounder of the art-science of linear perspective, the humanist scholar Leon Battista Alberti (1406–72), stepped literally into the picture in 1435. In that year and the next, he wrote two versions of a treatise on painting, entitled *De Pictura* in Latin. The first edition was in the vernacular Tuscan dialect, which he rewrote several times in the classical language of Cicero, then the "universal" language of scholarship and learning everywhere in Western Europe much as English is becoming the universal language today.[5] This famous work became the first ever to treat the visual arts as an appropriate humanist subject, as worthy of the same intellectual study as the great classics of antique Greek and Roman literature.

Alberti, who certainly accepted without question all the religious analogies between perspective optics and divine intention, nevertheless preferred to bring the matter more down to earth, as it were. Indeed, he may have been motivated by the complaints of certain contemporaneous painters in Florence that attempting to depict sacred space with the mundane rules of geometry was somehow irreverent. Even if Brunelleschi's mirror demonstration should fail to reveal divine order convincingly, Alberti realized that the very logic of Euclid's optical ordering must in any case signify human moral ordering. The humanist scholar, only recently arrived in Florence, was so taken by the visual arts flourishing in that fecund city that he even dedicated the vernacular version of his book to Brunelleschi. In fact, he now believed that painting in particular, if it followed Brunelleschi's geometrical structure correctly, could encourage ethical human behavior just as surely as the writings of Plato and Aristotle.

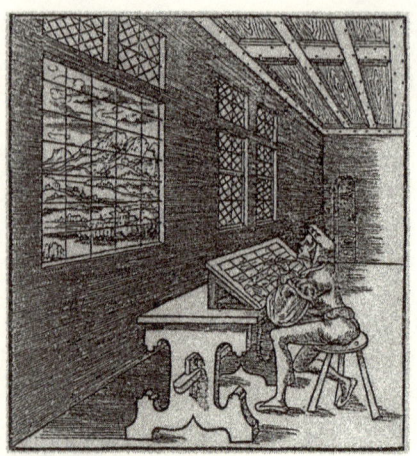

5. *Alberti's Window.* Woodcut from Johann II of Bavaria and Hieronymus Rodler, *Ein schön büchlein und unterweisung der Kunst des Messens*, Simmern, Germany, 1531.

Alberti's perspective method as described in his treatise was certainly a codification of Brunelleschi's method already in practice by a number of artists by 1435, but he did present it in the form of simple sequential steps which, as his treatise increasingly circulated in Italy and across the Alps, helped to proliferate the new art-science throughout Europe. His most original contribution, however, has ever since become known as "Alberti's window" (ill. 5), to be described and discussed in later pages. Its importance, in essence (even if originally unintended), is that it subtly shifted the object of perspective painting away from "mirroring" Nature as if it were a mere reflection of God's true brilliance in heaven, to seeing Nature instead as if through an open window, not as a divine mystery revealed by geometry, but as worldly perfection framed by geometry.

By the early sixteenth century, however, even as Alberti's "window" was accepted almost everywhere in Western Europe as providing the ultimate illusion of visual reality in art, Italian painters, while not abjuring the optical truthfulness of Alberti's perspective, nonetheless began to tire of creating "window" illusions of deep space, and, ever looking for new trompe l'oeil effects, to become excited instead by the visual illusion of frontal projection. This new fascination was remarkably encouraged by recent archaeological discoveries of ancient Roman relief sculpture, where figures were carved protruding from the surface of stone or plaster, lined side by side as if in lateral procession with their raised forms made visible by sharp contrasts between lighted and shaded sides. Instead of simulating a view of deep space beyond the pictorial surface, the ancient carvers created an equally lifelike simulation of forward projection, further emphasized by the shadows their forms cast against the background plane. This new archaeological fascination, especially after the 1520s, resulted in a widely popular "relief-like" style of classical painting in central Italy, as Marcia Hall has termed it. Alberti, of course, had already urged painters to achieve such mastery, and his perspective rules concerning the importance of manipulating black and white in order to create

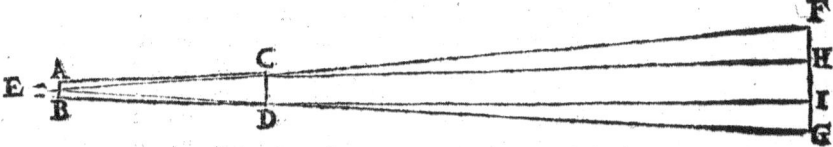

6. Galileo, engraving showing the optical geometry of his telescope, *Sidereus nuncius*, Venice, 1610. Courtesy of Jay M. Pasachoff, Williamstown, Massachusetts.

the illusion of shadows and raised relief were still fundamental to the new style, especially after the first printed publication of his treatise on painting in 1540.[6]

Finally, I look in upon seventeenth-century Florence, nearly two hundred years after Brunelleschi's mirror and Alberti's window have impressed their profound effects upon European art and thought. Here we encounter Galileo Galilei (1564–1642), the great astronomer and physicist, and by no coincidence a direct descendant of his native city's illustrious artistic tradition. In fact, Galileo was both a talented draftsman and a teacher of perspective drawing and shadow rendering. One of the lessons he studied and taught was how to draw a sphere with raised protuberances casting shadows on its surface in raking light.

During 1609 and 1610, Galileo built himself a telescope, based on news of its prior invention in the Netherlands where the novel instrument had attracted attention for its military possibilities. The geometry of its function, as Galileo himself diagrammed it later (ill. 6), was essentially based on the same Euclidian optical model as Alberti's perspective: a "visual triangle" with fixed eye-point at E, object of sight at FG, and intersecting "window" at CD (Galileo, of course, added magnifying lenses in between at AB and CD, refracting the incoming light rays to bring the object of sight appear much closer at HI).

In England, Thomas Hariot, a renowned mathematician and explorer, had already procured a version of the new device that he appropriately

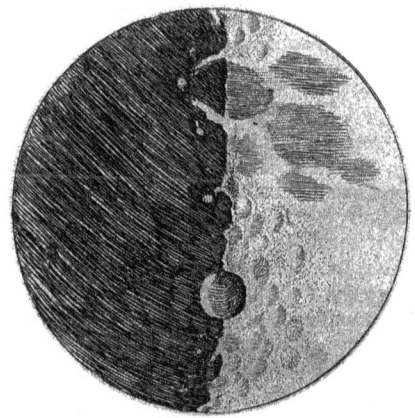

7. Galileo, engraving of the moon, *Sidereus nuncius*, Venice, 1610. Courtesy of Jay M. Pasachoff, Williamstown, Massachusetts.

called a "perspective tube." He even used it to look at the moon, but he still saw nothing to change his mind that the lunar body was anything other than a perfect sphere, yet he did notice its "strange spottednesse" but had no idea as to the cause.

However, when Galileo also aimed his modified instrument at the moon, his unique drawing and teaching experience made it clear to him that Hariot's "strange spottednesse" was really caused by dark shadows cast by protruding mountains on the moon's irregular surface. To the startled public who read his book, *Sidereus nuncius* (Starry Messenger), which he rushed into print in 1610, Galileo's version of Alberti's window quite shattered Brunelleschi's mirror. What Galileo indeed revealed was that the earth was not necessarily a pale reflection of the immaculate heavens, as Brunelleschi's mirror had proclaimed, but in the case of the moon just the other way around. Beyond even the most religious zealot's doubts—that is, if one of them dared to look through his "perspective tube"—Galileo proved that the first "planet" in Dante's magnificent ascent to the heavenly empyrean was hardly the "eternal pearl" described by the poet, but was rather a most imperfect sphere, marred and crinkled just like the lowly earth.

CHAPTER TWO

Crisis in Christendom

Beside the principal doorway of the Florentine Church of Orsanmichele, in a niche looking out upon the main street connecting the secular governmental headquarters, the Palazzo Signoria (now referred to as the Palazzo Vecchio or "Old Palace"), and the most important religious shrine in the city, the Cathedral (Duomo) of Santa Maria del Fiore, is a lifesize bronze

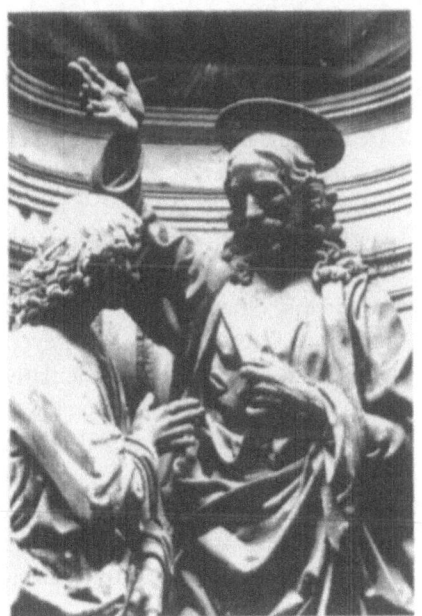

sculpture of Saint Thomas the Apostle, inquisitively thrusting his fingers toward the gash where Longinus's lance pierced the side of Jesus side as he expired on the cross (ill. 8).

Notice that Jesus is deliberately parting his garment to frame the gaping wound almost as if he were presenting it between the curtains of a theatrical stage. Thomas stares in wonder, extending his quivering fingers as if to feel it tactilely and thus prove sensately that Jesus truly had returned from the dead.

8. Andrea Verrocchio, *Christ and Saint Thomas*, Church of Orsanmichele, Florence, 1467. Photo: Gabinetto Fotografico del Polo Museale Fiorentino

The sculptor of this remarkable work, Andrea del Verrocchio (ca. 1435–88), whose very name means "true eye" and who in 1467, the year of its commission, had just taken on a young apprentice named Leonardo da Vinci, has here vividly demonstrated what I consider a profoundly new perceptual awareness, which by now late in the century had quite altered the collective sensitivities of all art-loving Florentines, artists and viewers alike.

Whether or not Verrocchio realized the artistic implication of this famous biblical test, his eloquent rendition of the episode also proclaimed that artists, as well as divine beings, could no longer depend on two-dimensional symbolic shapes in order to have observers believe in the "reality" of their imagery, but rather must create a sense that whatever is represented, even miracles, should appear to occur in volumetric space figuratively tangible from every side. What caused this new perceptual insistence? What specifically had inspired artists, especially in Italy, to so dramatically reform their art in this direction?

Although the major event of this shift occurred earlier in the fifteenth century, there was certainly a long run-up that can be traced all the way back to the seventh century, and even to the glorious days of the Roman Empire or at least to its nostalgic memory as it began to be revived in Christian Europe in the following Middle Ages. Imagine yourself a devout Christian living in Western Europe during those medieval times, pondering how you might have reacted to a pictorial *mappamundi* or "map of the world," diagrammed in the then traditional two-dimensional "T-and-O" format (ill. 9).[1]

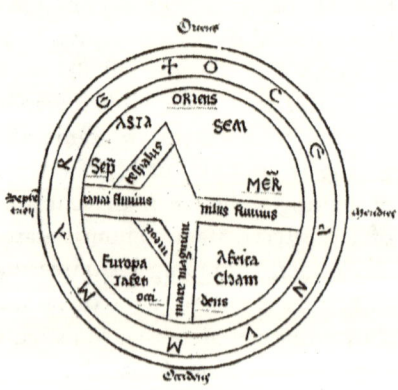

In your indoctrinated mind, you would take for granted that it shows the original master plan of sacred Earth as conceived in the eye of God at Genesis. While the flattened shape of this imaged "world" is circular like the letter "O," it did indicate that the Earth is indeed a sphere, just as the ancient Greek philosophers had long since demonstrated. However, you would not yet be particularly concerned that the geometric curvature of a three-dimensional globe might affect and distort whatever is seen on its surface, especially from an implied bird's-eye viewpoint you have unconsciously assumed was God's.

9. Woodcut of a T-and-O map from Saint Isidore of Seville, *Etymologiarum sive originum libri XX*, Strasbourg, 1473. Chapin Library of Rare Books, Williams College, Williamstown, Massachusetts.

In any case, the earthly surface as diagrammed reveals only three schematically conjoined land masses representing Asia, Europe, and Africa surrounded by a vast "Ocean-Sea" which presumably continues under the invisible bottom side of the world-sphere. These three known continents (the Americas and the polar regions, of course, were undiscovered) are arranged with the direction east uppermost. Asia therefore fills the whole top half of the "O." At the furthest eastern edge of God's Creation, just where the Bible says it is, you know there must be located the Garden of Eden where God made Adam and Eve.

Two connected horizontal bands, representing the natural rivers Don and Nile, separate Asia to the upper east from Europe and Africa on the lower western side. Europe and Africa in turn are separated north from south by a vertical band representing the Mediterranean Sea. Together these bands of water form a "T," reminiscent of Jesus's cross superimposed on the "O." Just above the crossing of the "T," at the furthest western side of Asia but in the very middle of the map, the city of Jerusalem is assumed to be positioned, just as God deliberately "certified" this place by linking his divine eye on axis with the very center of the Earth's visible surface, thus appropriating this location for the prophesied birth and sacrifice of his son Jesus Christ. As the Bible testifies, "Thus sayeth the Lord God: this is Jerusalem: I have set it in the midst of the nations and countries that are round about her" (Ezekiel 5:5). In later pages of this book describing the science of optics it will be shown just how relevant this traditional cartographic assumption was to your medieval understanding of how God envisioned the universe at Genesis according to the natural laws of Euclidian optics.

Indeed, in the understanding of sacred "reality" during the Middle Ages, it was no coincidence that God conceived of the Earth as consisting of only three continents. Obviously, he wished to signify that most mystic of metaphysical numbers, alluding to the eventual revelation of the holy Trinity whose carnal embodiment in Jesus on his "T"-shaped cross saved all Mankind. Thus, the map shows the lands of the world not only literally as God formed them but tropologically, in preparation for the birth and sacrifice of the Savior, and the eventual Apocalypse and Last Judgment.

Furthermore, the unique position of Jerusalem almost at the juncture of the three traditional continents even by modern cartographical reckoning was again no haphazard coincidence. Both geographically and tropologically, it was consigned by God to be his unique *umbilicus mundi*, on direct perpendicular axis with heaven, and thus for mortal Christians the holiest of holiest shrines wherein to pray for your promised redemption.

But alas, your Christian Holy Land was now occupied by the infidel believers in another divine prophet named Mohammed. From the seventh century on, they had most unexpectedly and incredibly conquered and converted not only all of western Asia and northern Africa but were even surging into Europe through Spain and the Balkans. Nevertheless, it really wasn't until the eleventh century, when the occupying Mohammedans ordered the destruction of Christian shrines and prohibited Christian pilgrims from visiting Jerusalem, that Europeans felt the true horror of what they had lost: the very psychological wellspring of their own religion.

Hence, Pope Urban II, pastor of all Christendom, set out in 1095 to rally the disparate kingdoms of Europe and launch the first (of yet seven more) Crusades, but all in vain as it eventually turned out. In 1099 Jerusalem was in fact retaken, remaining Christian for nearly a century, then was lost again in 1187, reoccupied once more in 1229, but only intermittently held until 1291 when the Egyptian Mamluks drove the exhausted Christian forces out the Holy Land for good.

For the next two hundred years or more, Catholic Europe was not only without easy access to its sacred umbilical center but it was ravaged by other apocalyptic catastrophes, like the distressful schism in the Church when the papacy itself was torn apart by national rivalries, the devastating bubonic plague in which nearly a quarter of the population of Europe perished, and the unnerving fear of militant, advancing Islam during the hundred-year siege of Constantinople and its traumatic fall to the Turks in 1453 (comparable to the crisis a Soviet seizure of Berlin might have caused had it happened during the cold war). It was as if a wrathful God was deliberately wreaking revenge on his own chosen people for their disgraceful surrender of Jerusalem.

What irony, as you pondered that glaring T-and-O map, that the most sacred shrines of every other great civilization including Islam were safely protected within their own ethnic and ideological confines. Indeed, you would have discovered to your further disappointment that contemporary Mohammedan maps showed Arabia and Mecca in the center, and that Chinese charts similarly positioned the imperial city of Beijing. How humiliating to be so blatantly reminded that your own holiest of Christian centers was in fact "off center" from your own political and religious control.

Nevertheless, even though this cartographic anomaly was a psychological embarrassment, it also served as an equally powerful stimulant to your traumatized faith, to reinvigorate it with new ideas, encouraging you to learn and apply more science and technology in order to thwart advancing Islam and even retake Jerusalem. It was almost as if the T-and-O map was a direct order by God himself for you to spread his word to those heathen lands surrounding Jerusalem, and convert them by whatever means to his one true religion, thus restoring your holiest city to its originally ordained place, the center of an all-Christian world.

No one articulated this desire more eloquently during the thirteenth century than the English Franciscan Roger Bacon (ca. 1220–92). His principal treatise, the *Opus majus* (Major Work), written in the 1260s, is replete with calls upon Christian leaders to study the *quadrivium*, the four mathematical sciences among the seven classical *Artes Liberales* (Liberal Arts), including most especially Euclidian geometry and its optical subscience, the theory of vision and light.

For instance, Bacon described in a separate section entitled "On the value of optical marvels in converting the infidel" how one might apply the optical fact that light rays reflected from a concave mirror can cause fire because of their focused perpendicular strength. Thus, giant "burning" mirrors might be manufactured and deployed in warfare to burn the enemy's cities.[2] Bacon added further that the same new mathematical sciences might have application to the visual arts, musing that basic geometry even applied to religious images might serve to renew enthusiasm for the next Crusade.

Oh, how the ineffable beauty of divine wisdom would shine and infinite benefit overflow, if these matters relating to geometry, which are in Scripture, should be placed before our eyes in their physical forms! For thus the evil of the world would be destroyed by a deluge of grace. . . . Surely, the mere vision perceptible to our senses would be beautiful, but more beautiful since we should see in our presence the form of our truth, but most beautiful since aroused by the visible instruments, we should rejoice in contemplating a spiritual and literal meaning of Scripture because of our knowledge that all things are now complete in the church of God, the bodies themselves sensible to our eyes would reveal. Therefore, I count nothing more fitting for a man diligent in the study of God's wisdom than the exhibition of geometrical forms of this kind before his eyes. Oh, that the Lord may command that these things be done! There are three or four men who would be equal to the task, but they are most expert among the Latins; and rightly must they be expert, since unspeakable difficulty lurks here, owing to the obscurity of the sacred text and contradictions of the sacred writers and difference of the other expounders.[3]

Who might Bacon have meant by the "three or four experts among the Latins"? He was supposedly writing this in Paris during the early 1260s, but could he have previously traveled to Italy and there heard of, or admired firsthand, some early examples of the "Roman School"? Here practiced a number of painters and mosaicists whose renascent antique classical style was to inspire the revolutionary frescoes in the new Basilica of Saint Francis at Assisi, his own mother church, thus setting the stage for Giotto di Bondone of Florence (1266/7/76–1337), and the birth of the Italian Renaissance.[4]

Or could Bacon also have been referring to another "Latin" art form: ancient Roman *scenographia* (the design and construction of theatrical sets)? Perhaps he was advocating that this long forgotten antique subscience of geometry likewise be applied to the staging of "miracle plays" that were gaining popularity everywhere in medieval Europe.[5] Such attractive but often naïvely awkward presentations were being performed during all the major religious holidays of the Christian year, dramatized by costumed actors or three-dimensional mannequins on makeshift and often mobile stages in the streets and public squares of all the towns of Europe. Indeed, Bacon in his wide reading of classical sources may have been intrigued by something like the following comment from the first century BC Roman architect Vitruvius,[6] describing, for the benefit of his own contemporaneous theater designers, how the Greeks before him created illusionistic backdrops by means of what must have been a prescient early form of geometric linear perspective:

> *Scenographia* is also the shading of the front and the retreating sides, and the correspondence of all lines to the center of a circle. . . . For to begin with: Agatharcus at Athens, when Aeschylus was presenting a tragedy, was in control of the stage, and wrote a commentary about it . . . in order to show how, if a fixed center is taken for the outward glance of the eyes and the projection of the radii, we must follow these lines in accordance with a natural law, such that from an uncertain object, uncertain images may give the appearance of buildings in the scenery of the stage, and how what is figured on vertical and plane surfaces can seem to recede in one part and project in another.[7]

To reconstruct how riveting these miracle plays must have been to medieval Christians obsessed with being eyewitnesses to the lives of their favorite sacred heroes and heroines, it will be useful to return vicariously to the small Italian hill town of Assisi and visit the birthplace of St. Francis, the mother convent of the Franciscan order and major sanctum for Christians ever hoping to discover God's divine presence in late medieval Europe. The great basilica honoring St. Francis in Assisi was begun and completed in the thirteenth century as a sepulcher for the town's favorite son who had been canonized in 1228, only two years after his death. Within months after confirmation of St. Francis's miraculous stigmata had spread through Europe, thousands of pilgrims flocked to the new shrine in order to touch his tomb, to see his relics, and, perhaps just as important, to re-create the events of his life for themselves as street performers in public *laude* (the Italian name for such spectacles).

To stimulate further these popular responses, the interiors of the upper and lower churches of the two-tiered basilica were painted from top to bottom with fresco cycles narrating Christian stories and theological themes by various Italian artists. The most famous cycle is that of the *Life of St. Francis*, painted at the very end of the thirteenth century. Twenty-eight scenes follow in a left-to-right sequence around both sides of the lower nave in the upper church, to be viewed like a modern comic strip. They are arranged in groups of three on each side of the first three bay walls, and in groups of five on the two sides of the entrance bay. Each event of the saint's life is depicted as if it were an individual stage set of a miracle play, and all are framed together under a fictive modillion cornice that runs around both sides of the church, very much reminiscent of a real cornice that covers the colonnaded portal of an ancient Roman temple still standing just up the street and converted into a Christian shrine named Santa Maria Sopra Minerva.

In the very first painted scene of St. Francis's atavistic life (ill. 10), we see a citizen of Assisi laying down his cloak before the young charismatic, right in front of that very Santa Maria Sopra Minerva. What we have in this fresco, I believe, is not so much an imagined recapitulation of what happened to St. Francis in actual life as it is what the artist and contemporaneous viewers would have witnessed in a theatrical representation of the event performed in the public square before this familiar local landmark.

What the Assisi painter ingeniously did was to apply the colonnaded motif of Santa Maria Sopra Minerva's classical facade as a framework for all the painted scenes, reminding viewers that each was really a *tableau vivant* taking place before that building in Assisi's main piazza. Furthermore, the painter of this proscenium-like architectural scheme, perhaps Giotto himself, would seem to have been familiar with the very Greek and Roman theories of stage design mentioned earlier.

Look closely now at illustration 11 and notice how the projecting modillions of the painted cornice are angled left and right so that they appear illusionistically to converge over the center of each group of three scenes. The optical convergence at this point compels the viewer to stand fixed in the middle of the bay and observe the scenes as if they were moving sequentially behind a single colonnaded proscenium. Art historian Charles Parkhurst has further shown that the depicted background architecture in the early fourteenth-century frescoes narrating the *Life of the Virgin and Christ*, this time surely painted by Giotto in the Arena Chapel, Padua, a few years later,

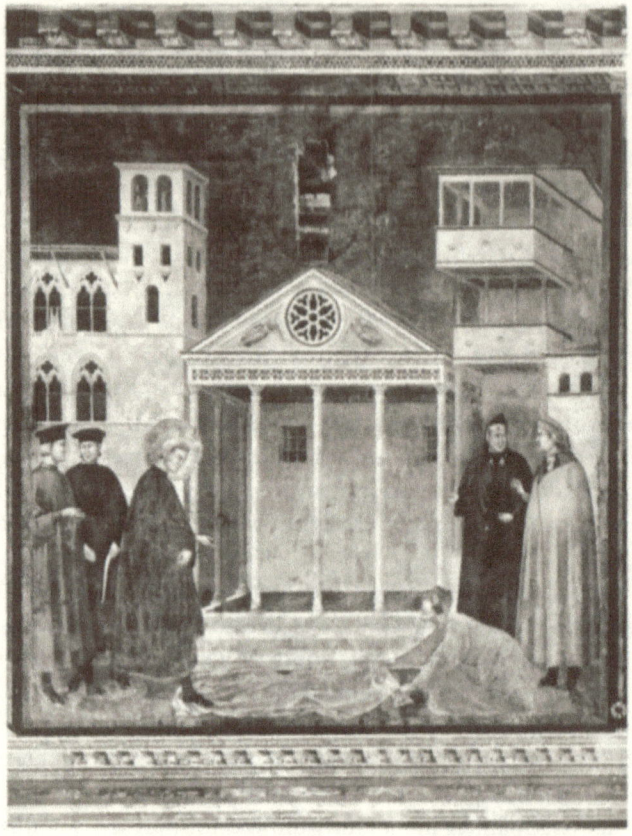

10. Scene one from the life of St. Francis, from bay A, right wall,
upper church, Basilica of St. Francis, Assisi, late thirteenth century.
Photo: Alinari/Art Resource, N.Y.

actually represents assembled stage props that could be reconverted into different settings from one narrative scene to the next (ill. 12).

In other words, the mural paintings of both Assisi and the Arena Chapel,
long considered as the formative models of all subsequent narrative representation in Western Renaissance art, seem to have been motivated not so much
by attempts to represent the illusion of "real life" but rather to re-create the
illusion of stage settings for contemporaneous miracle plays.

Inevitably, this peculiar urge to have two-dimensional pictures ape the
illusion of three-dimensional, tactile reality, especially prompted as it was by
the increased vividness of staged miracle plays, led artists in Christian Europe

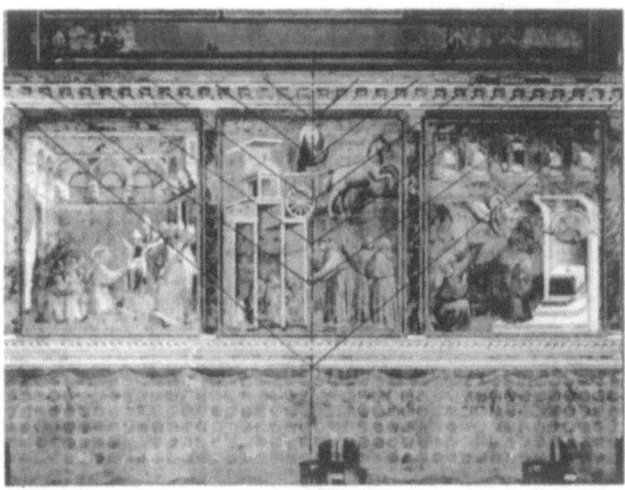

11. Perspective reconstruction by author of bay C, right wall, Basilica of St. Francis, Assisi, late thirteenth century.

to explore various other technical and mechanical skills from whatever sources were then available, which they might adapt to their art. One such was surely the science of optics, which will be discussed in chapter 3.

Bacon, of course, had no inkling that he might also be adumbrating the Italian Renaissance notion of linear perspective in the visual arts, and thus the association of a painting with what one sees reflected in a mirror or

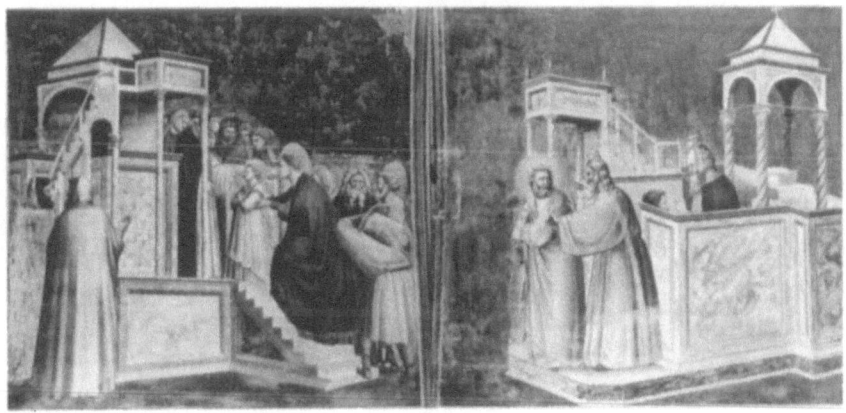

12. Giotto, scenes from the life of the Virgin Mary, Arena Chapel, Padua, ca. 1305. Photo montage by author.

through a window. Nevertheless, this remarkable geometric adaptation, first demonstrated in Florence, Italy, by Filippo Brunelleschi around 1425, and then codified by Leon Battista Alberti in his *De Pictura* of 1435/6, stemmed not initially from any deterministic premonition of secular science, but rather from the longing of medieval Christians to feel that God and his holy works be more palpably present and immanently concerned with the problems of their daily lives, assuaging their feelings of spiritual emptiness caused by the loss of Jerusalem.

In a nutshell, the original inspiration for linear perspective was a very medieval solution to a very medieval problem. No doubt for this reason, the very first monumental work of art in true geometric linear perspective after Brunelleschi's initial demonstration was not some secular subject, as we today might expect, but the most esoteric spiritual mystery in all Christian doctrine, the Holy Trinity, as painted by Brunelleschi's young friend, Masaccio (1401–28?), in 1425, as will be examined in detail in chapter 9.

CHAPTER THREE

And God Said, "Let There Be Light!"

Perspectiva naturalis

We speak now of the spread in Western Europe of the science of ὀπτική, "optics," formulated long before the Christian era by the ancient Greeks, and based on the fundamental laws of plane and solid geometry as first defined by Euclid. He and other Greek philosophers then reasoned that human vision could be explained by these same geometric laws.

As the Roman Empire fell apart in the early Middle Ages, the great heritage of classical science remained largely inert in Europe. Meanwhile, between the seventh and eleventh centuries AD, while Mohammedanism was sweeping across the ancient Alexandrian world, Muslim savants rediscovered this rich Euclidian tradition with special admiration. They translated many of the Greek mathematics and optics texts into Arabic, even adding their own commentary. Only after the Christian reconquest of Moorish Sicily and northern Spain in the twelfth century were several of the old classical optics texts, heretofore forgotten in the West, recovered, but mostly in Arabic renditions. Patiently, *verbum ex verbo* through the next two hundred years, these manuscripts were retranslated into Latin, and slowly circulated among the monasteries and universities of Europe, inspiring a number of new treatises on the same subject. Two of these were written by English Franciscans, Roger Bacon and John Pecham (ca. 1235–92). A third important work was composed by a contemporaneous Silesian scholar with Dominican connections named Witelo (fl. 1260s).[1]

The most influential Arabic commentary, inspiring especially the European authors just cited, was composed by Abū ʿAlī Al-Hasan Ibn Al-Haytham (ca. 965–ca. 1041) known simply as Alhacen in the West, whose treatise, originally called *Kitāb al-manāzir* or "Book of Optics," became known as *De aspectibus* or *De perspectiva* in its Latinized editions.[2] *Perspectiva* was a recently minted Latin word deriving from *perspectus*, participle of *perspicere*, meaning "to see through."[3] Sometimes qualified as *perspectiva communis* or *perspectiva naturalis*, it soon became the accepted name for the revived science of optics everywhere in medieval Europe.[4]

Optics, under whatever linguistic title, originally had only to do with explaining the nature of light rays, how they always travel in straight lines, how they are reflected in mirrors, how they are refracted when striking obliquely or entering a denser medium, and, most of all how they affect the way the human eye sees. This last point should be stressed. As David Lindberg has emphasized, "The essential character of medieval optics, distinguishing it from modern optics, is that there is no optical problem without an observer. Medieval optics is a theory of vision."[5]

It should further be emphasized that there is nothing in original *perspectiva* theory that had anything to do with the visual arts. Furthermore, painting and sculpture did not become an "optical problem with an observer" in the full geometric sense until the fifteenth century when the Florentine artisan, Filippo de Ser Brunelleschi, and the Florentine humanist, Leon Battista Alberti, adapted *perspectiva communis* to what later would be called *perspectiva artificialis*, linear perspective for painting.

To be sure, original *perspectiva communis* was regarded from the beginning as the special handmaiden of Euclidian geometry. The latter branch of mathematics had also been taken over from the Greeks by the Arabs, and only revealed again in the West after the same reconquest. Since light rays were understood by the ancients as radiating from every point on the surfaces of whatever their source, whether a tiny candle flame or the gigantic sun, they always flare outward in conical or pyramidal sprays (the two terms, *conus* and *pyramis*, tended to be interchangeable, and meant the same thing in the medieval optics texts). Similarly, light rays in the same cone/pyramid form stimulate vision by entering the eye through its apex end. (ill. 13).

It was thus reasoned that visual images were "framed" at the base of the visual cone/pyramid (let us assume "visual pyramid" hereafter) and as they entered the eye must conform to Euclid's fundamental law of similar triangles. For instance, in any triangle (ill. 14) with A as apex, altitude AC, and base BCD, if A is thought of as the point of sight, and base BCD the surface seen, then the same proportion as between the distance AC and the base prevails between the distance from A to all the parallel intersections within the same extending triangle, such as EFG, HIJ and KLM. In other words AC:BCD as AF:EFG as AI:HIJ as AL:KLM, and so forth.

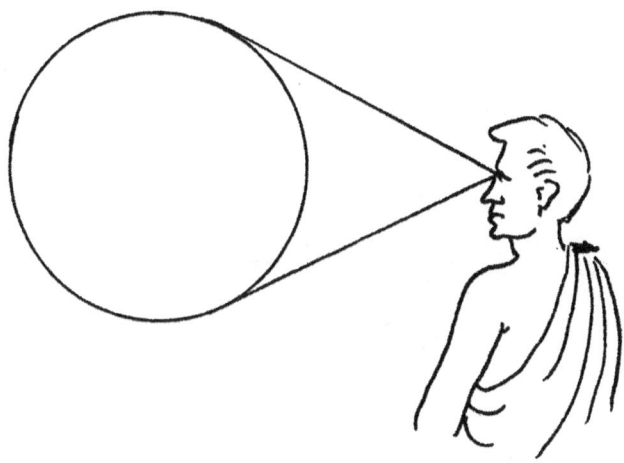

13. The visual cone/pyramid. Drawing by author.

Greek and Arab commentators on Euclidian optics were quick to realize the significance of this in explaining how images of very large objects can decrease in size in proportion to their distance from the viewer's eyes, and thus be able to penetrate the tiny pupil of the eye. For instance, in the same diagram (ill. 15), let A again stand for the human eye, and BCD the object being observed (Arab commentators liked to use the "camel" as their example).

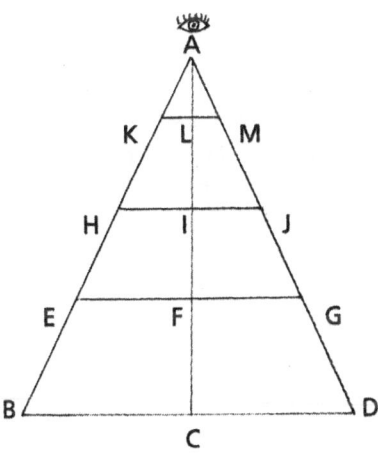

14. The visual triangle. Drawing by author.

As the distance AC between these points diminishes to AF and then to AI, AL, and so on, the illuminated "camel" will grow ever smaller in proportion until it is finally able to fit through the pupil and be seen. Much argument occurred at first as to whether this mode of "seeing" was energized by light coming from the camel into the eye (intromission) or whether some force had first to be projected from the eye (extromission) fixing on the camel before it could be made sensible. Although Alhacen pretty much proved that intromission was the more logical theory as

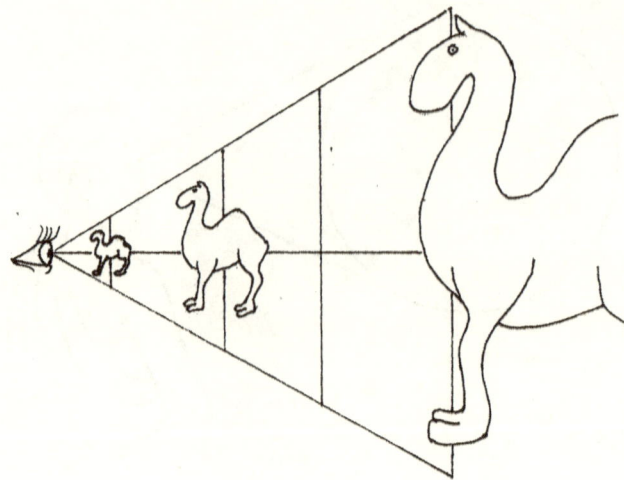

15. How a "camel is seen" via the visual triangle. Drawing by author.

far as eyesight was concerned, the movement of light rays in either direction could be described by the same geometry. Nevertheless, this simple geometric fact, that light rays can be diagrammed as straight lines, extending into or out from the eyes, may have provided the most critical scientific rationale for explaining how the Christian God spread his divine grace throughout the entire universe and thus interacted with mankind.

Just as interesting to medieval perspectivists was the optical theory of mirrors, which was becoming a special subscience called catoptrics. Mirrors, of course, have fascinated mankind since the earliest Neanderthal Narcissus was amazed at recognizing his own face reflected in a pool of water. Moreover, mirrors were always suspected of having magical power, allowing one to access the spirit world as if through a mystic portal. Even during the very Christian Middle Ages in Europe, it was popularly believed that if a mirror reflected the image of a holy relic, the latter's miraculous potency would also be virtually captured. Pilgrims often carried small round mirrors to sacred shrines in order to reflect the relics. The mirror thus metaphysically "charged" became a surrogate talisman in its own right.[6] No doubt, this primordial allure of mirrors encouraged catoptric investigation, holding out the possibility that the science of geometry would penetrate and rationalize the mirror's heretofore occult mystery.

What the Greeks had discovered was that the mirror reflections of light rays and illuminated objects also conformed to the laws of Euclid. For in-

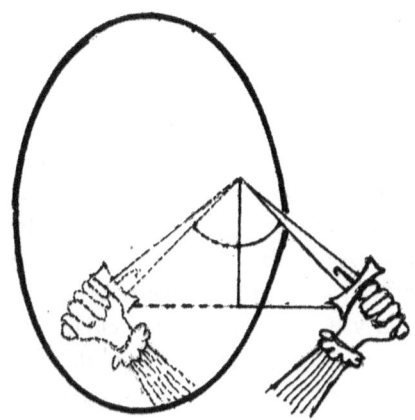

16. Mirror angles of incidence and reflection. Drawing by author.

stance, an object at an angle before a mirror is always reflected at the same angle (ill. 16: the angle of reflection always being equal to the angle of incidence).

By the same rule, a light ray falling perpendicular upon the mirror surface appears in the latter's virtual space to be reflected directly back upon itself. Similarly, the reflection of an illuminated object perpendicular to the mirror surface will appear always on the same perpendicular line and just as far within the virtual space as the object is before it, but with left and right reversed.

Furthermore, medieval optical commentators regarded objects reflected in a mirror directly opposite, on the same perpendicular, as more distinct, thus "stronger" at the mirror intersection. The single ray in the center of any pyramidal spray of light rays either falling perpendicularly on a mirror surface or emanating from the center of a light source was called the *axis visualis* (ill. 17) or just as commonly the *axis perpendicularis* ("perpendicular axis").

On the other hand, when light rays pass from one medium into another, as for example from air into water, they are refracted or bent, and thus bear a "weaker" image.

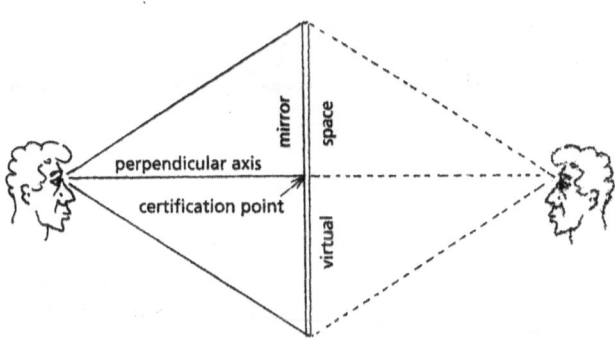

17. The *axis perpendicularis* as reflected in a mirror. Drawing by author.

Finally, the physiological structure of the eye itself was believed to behave like a mirror. The convex lens within the eye, then called the *crystallinus*, was understood to receive convergent visual rays from the observed object and display the latter's minutely scaled image on its surface just as if it were being reflected in a mirror, which image then passed through the *crystallinus* to the optic nerve in the back of the eye for ultimate cognition in the brain. Furthermore, the same perpendicular axis principle obtained in the visual pyramid, extending from the center of the object being perceived to the center of the *crystallinus*, thus "certifying" (*certificandum*) that point in the object that is to be discerned most clearly. Only the perpendicular axis passed through the *crystallinus* straight and unrefracted: thus, if the eye wishes to see the whole of a large object just as clearly, it must move its perpendicular-axis focus from point to point across the object's surface. As the perspectivists described it: "Perception is certified by the axis being conveyed over the visual object" (*apprehensionis certificatio per axem [perpendicularem] super visibile transportatam*).[7]

Again by the same rule, all oblique rays in the pyramid coming to the eyes from the object are refracted as they pass through the denser *crystallinus* medium, thus transmitting weaker image details to the brain. Long before Johannes Kepler discovered that such refraction actually caused the image to invert inside the eye (how the brain then managed to revert it once again, he

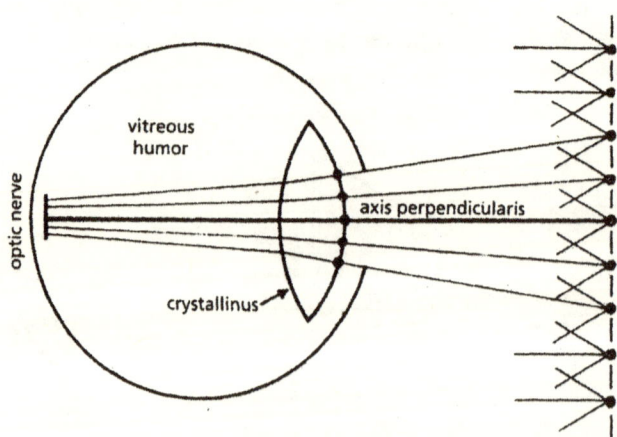

18. The medieval notion of how light enters and is refracted inside the eye. Drawing by author.

was never able to explain), my edited diagram (ill. 18) shows how earlier medieval philosophers erroneously interpreted the way oblique visual rays, refracted while passing through the *crystallinus*, might then continue parallel with the *axis perpendicularis*, thus maintaining the image upright to the optic nerve.

Medieval Christian theologians were fascinated by these Greek and Arab revelations. The Latin word for mirror, *speculum*, became almost a synonym for "divine revelation." Numbers of treatises with titles such as *Speculum naturae* (Mirror of Nature) or *Speculum salvationis humanis* (Mirror of Human Salvation) were published throughout medieval Europe. Furthermore, manufacturing technology was gradually improving. By the beginning of the fifteenth century both convex mirrors of glass and flat mirrors of silvered metal were produced in convenient size, more and more becoming standard household items where their reflections began to be compared, both actually and symbolically, with painted pictures.[8]

The English bishop Robert Grosseteste (ca. 1168–1253) noted further in his own writings that the reason God created light on the first day of Genesis was that he needed an essential medium through which his every action must travel between heaven and earth, especially the spreading of his divine grace to mankind.[9] Since luminous bodies propagate light in linear rays in all directions from single points, Euclid's theorems, inscribed in straight lines like light rays on uniform white paper like the illuminated atmosphere, could explain exactly how this process worked geometrically in God's divine mind, literally providing the key to his intelligence, and how he conceived and projected the universe from its minute conception in his mind's eye to its full-blown realization in the void. Christian philosophers were likewise able to apply the same process to explain how God extends his moral message to the human soul. As the German mystic Meister Eckhart (1260?–1328) reasoned, the soul "sees" God just as the eye receives light.[10] Moreover, the soul reacts like the *crystallinus* inside the eye, mirroring God's grace but only when "certified" (*certificata*) by the metaphysical perpendicular axis traveling from the center of God's divine mind to the center of the immaculate soul. If, however, the soul is stained with sin and not immaculate (like a cataracted lens or a clouded mirror), God's grace must thus be refracted away. Even the Tuscan poet Dante Alighieri took note of the power of the perpendicular axis, as he remarked in his *Convivio*:

It should be known that, although more things than one can enter the eye at one time, that which comes in a straight line into the point of the pupil is

truly seen by it, and alone stamped on the imagination. And this is so because
the nerve along which the visual spirit travels runs straight to that part, and,
therefore, in truth one eye can not look at another without being seen by it,
because as the eye which beholds receives the form on the pupil along a
straight line, so also its own form goes along by this same line into the eye
which it beholds.[11]

In effect, Dante implied, you can't tell an honest man unless you look him
straight in the eye.

Members of the young Franciscan order, headquartered in the basilica
dedicated to their founding saint in Assisi, Italy, were especially moved by
such optical and catoptrical analogies, none more so than the Englishman
Roger Bacon. His most imaginative application of *perspectiva* geometry, however, was to hypothesize that every physical and metaphysical object in the
universe gives forth from itself a uniform but invisible force that he termed
species, the Latin word for "likeness" or "form," but which in his refined definition acts exactly as light rays in the visual pyramid, with the perpendicular
axis as the most powerful. *Speciei* (pl.) then "multiply" (*multiplicant*)—they fan
outward from every point on every object and then spread through the universe, interacting with *speciei* from other objects. Some sources produce a
stronger *species*, termed by Bacon as *agens* ("agent"), which can render a
qualitative change on a certain other *species*, called *patiens* ("patient"), from
relatively weaker sources. For example, the agent *species* from the sun over-
powers the patient *species* of ice, causing it to melt.

Bacon's theory (which provided the title to his principal optical treatise,
De multiplicatione specierum) was thus quite concerned with how *species* would
actually pass through a medium-filled space, whether uniform and continuous like air, or from a metaphysical to a physical medium as between heaven
and earth:

> But a *species* is not a body, nor is it moved as a whole from one place to another,
> but that which is produced [by the *agens*] in the first part of the air [or other
> mediums] is not separated from the part, since form can not be separated from
> the matter in which it is unless it should be mind; rather, it produces a likeness
> of itself in the second part of the air, and so on. Therefore, there is no change of
> place, but a generation multiplied through the different parts of the medium;
> nor is it a body which is generated there, but a corporeal form that does not
> have dimension of itself but is produced according to the air; and it is not pro-
> duced by a flow from the luminous body, but by a drawing forth out of the
> potentiality of the matter of the air.[12]

In sum, by the end of the thirteenth century Christian theologians everywhere in Western Europe began to believe that the new geometric science of *perspectiva* not only provided the key to how God spreads his divine grace to mankind, but how he conceived the universe itself in his divine mind's eye at Genesis.

CHAPTER FOUR

Fra Antonino

We fast forward directly to fifteenth-century Florence, and address two contextual questions in order to understand how and why Brunelleschi and Alberti were motivated to apply, for the first time in world history, the geometric laws of *perspectiva communis* to the art of painting. First, we need to know if any of the many Latin and/or vernacular Italian treatises on *perspectiva* written during the previous centuries were actually circulating in Florence in their lifetime. If so, the second question: Why might artists, especially in Florence, suddenly be inspired to investigate *perspectiva*?

If fourteenth-century manuscript holdings in present-day Florentine libraries are any indication, the answer to question one is certainly yes. The Medici Laurentian Library still possesses three Baconian manuscripts, two of the *Perspectiva* section of his *Opus majus*, and another of the *De multiplicatione specierum*. The Biblioteca Riccardiana owns two more fourteenth-century editions of Bacon's *Perspectiva*, and yet another copy of *De multiplicatione specierum* is in the San Marco Library, purchased originally in the fifteenth century by either Cosimo de' Medici or Antonino Pierozzi At the same time in the university of Florence, lectures were being delivered concerning a treatise called *Quaestiones super perspectivam* (Questions on Perspective) by Blasius of Parma (Biagio Pelacani, d. 1416; two fifteenth-century manuscripts are still in the Laurentian Library), and yet another such Florentine work, anonymous but written in Italian as *Della prospettiva*, was also circulating in the city during the 1420s.[1]

Furthermore, there is no more telling evidence than the documented life of artist Lorenzo Ghiberti (1378–1455). Late in his distinguished career,

Ghiberti wrote a three-part treatise, known as *I Commentarii* (The Commentaries). The first two parts combine a history of art with a self-glorifying autobiography, but the third is of most interest because it is a compilation of quotations from various *perspectiva* texts, all translated into Italian, including many from Bacon, whom he never names. Although these are disorganized and it is difficult to discern just why Ghiberti collected them, many if not most of the excerpts stress the perpendicular axis and the necessity of it being centered on the object seen in order to "certify" clear vision. Although the jumbled *Third Commentary* offers no specific connection between *perspectiva* and works of art, his or anyone else's, it does certainly indicate that this subject deeply concerned him. Moreover, it also reveals that Ghiberti, who rose from humble origins to become a respected Florentine artisan, wanted very much to be regarded as an intellectual, to be accepted in the same lofty circle of savants that was embracing Filippo Brunelleschi, architect of the great dome over the city's cathedral, and Ghiberti's archrival.

Indeed, the science of *perspectiva* seems to have been one of the exciting au courant issues engaging the attention of the Florentine intelligentsia at the time. The theological and moral implications of optics had been stressed everywhere in Christian Europe since the thirteenth century, especially by the Franciscans. It was even being adapted for sermonizing by preachers in popular manuals such as *De oculo morali* (Concerning the Moral Eye) by one Peter of Limoges, dean of the Faculty of Medicine at the University of Paris during the 1260s, at the same time Roger Bacon and John Pecham were also resident there.[2]

The science of optics was such a hot topic in fifteenth-century Florence that artists as well as intellectuals—good Christians and regular churchgoers all—began likewise to be persuaded to learn something about the science, both for improving their pictorial techniques and making decisions about relevant subject matter.

Among the most admired and learned intellectuals in Florence during the first half of the fifteenth century was the Dominican priest Antonino Pierozzi (1389–1459). From 1439 until 1444 he was the Prior of the Convent of San Marco, the favorite charity of the Medici family and residence of the painter Fra Angelico. In 1446 he was appointed Archbishop of all Florence, in which post he served until his death. Less than a century later, in 1523, he was canonized as Saint Antonine.

Fra ("Brother") Antonino (as I shall continue to address him by his Italian name and title) was apparently a charismatic orator. From his various pulpits, he delivered powerful sermons to rapt audiences, touching on current matters of public morality, commenting on marriage, Florentine banking and business practices, ostentatious female dress habits, and even the excessive artistic embellishments of the city's churches, which he feared weakened the didactic purpose of holy images. His preaching always had the intention of warning Florentines against the venal temptations of mercantile prosperity.

Toward the end of his life, the Archbishop wrote down in Latin all the ideas and subjects he had been preaching publicly in a set of four volumes entitled *Summa theologica*, organized in obvious homage to the Dominican heritage of Thomas Aquinas. For our purposes the most interesting aspect of this monumental work was its emphasis on the act of seeing with the eyes, and his use throughout of optical analogies and metaphors to emphasize his spiritual and moral messages. Although he never referred to optics by its common Latin name, *perspectiva*, there is no doubt Antonino was familiar not only with the work of Peter of Limoges but with the basic principles of the visual pyramid as applied by Bacon to his *species* theory.[3] Following Peter of Limoges, Antonino liked to improvise on the science, often reforming the perspectivists' abstruse mathematical explanations into picturesque theological and moral *exempla*. For instance, while he explained how God's grace "multiplies" through the universe just as do the light rays from the sun, able to penetrate any medium such as glass "without rupture" (*absque scissione*), and even enter through a "closed door" (as did the Angel Gabriel into the Virgin Mary's sanctuary), he suggested the following catoptrical analogy in order to show how God's grace, like light, exposes the sins and temptations hidden in the mind:

> [God's grace] scrutinizes one's whole life, and deplores not only large sins but recognizes the smallest inordinate thoughts and influences. An abundance of [God's] grace falls just like a ray from the sun, where in a particular place numerous atoms (*atomi*), that is, tiny dust-like particles (*pulvisculi*), appear to be reflecting back (*reverberante*). Where, however, the ray is not reflecting back, nothing such as this is apparent. Hence, where a ray of [God's] grace illuminates, the mind recognizes all it own numerous defects.[4]

The first part of his four-volume *Summa* consists of twenty sections, each of which covers a basic theological matter beginning with a discussion of the soul in earthly life, the intellect, free will, the soul after death, sin in all its forms, and finally on law in all its contexts. The third section, "On Intellectual Power" (*De potentia intellectiva*), examines particularly how the intellect is the "eye of the soul" (*oculus animae*) and what is required of it for "seeing well both materially and spiritually" (*De his, que requiruntur ad bene videndum corporaliter et spiritualiter*).[5] Three things are necessary. First, the intellectual eye must be positioned directly (*recte*) before what is to be intellected. Antonino uses the common Latin adverb *recte*, here, but he surely meant, as would have been taken for granted by his contemporaneous readers, "perpendicularly."[6]

Second, an "illuminated medium" (*medium illustratum*) is needed between the eye of the mind and the "object suitably colored" (*objectum congrue colora- tum*); that is, as in optical reality. As example of what he meant by this re- quirement, Antonino quoted from Scripture, Luke 10:23, where Jesus says to his privileged disciples, "Blessed are the eyes which see what you see. For I say to you, many prophets and kings wished to see what you see, but they did not see."[7]

A few paragraphs later he spoke again about the necessity of the "me- dium" between eye and object to be clear, pure, and continuous (*medium debet esse lucidum . . . purum . . . continuum*). "Medium," according to An- tonino, applied to all four of the traditional classical humors or elements of nature: earth, water, air, and fire. Each occupied its own zone—earth at the bottom with water, air, and fire respectively above. Earth as medium is opaque, but the other three are successively more transparent. Air and fire, being the highest, are therefore the most "lucid and pure." Fire, metaphor- ically also signifying love and charity, had its sanctioned zone just below heaven itself, and thus it was the purest of all. Antonino went on to relate these elements to the Four Evangelists: Matthew to earth, Mark to water, Luke to air, and John, who loved Christ the most, should naturally be the apostle of fire, traditionally symbolized by an eagle soaring high above the earth. In an aside sure to rouse nods of self-satisfaction from his Florentine audience, Antonino noted that St. Mark, patron of rival Venice, "is signi- fied by water [and] chooses to live in a city founded entirely on water, that is of the Venetians, and which is symbolized by a lion which suffers from the fever (*quartana* [malaria]). So much water (*humore*) nourishes the fever."[8]

Antonino made yet another revealing reference to his knowledge of op- tics. As an example of unclear spiritual seeing he used the term *conus*—there can be no doubt that he was referring to the visual cone (synonymous with "pyramid" in all medieval *perspectiva* literature)—in describing how the eye in the medium of air might try to peer into the zone of fire, the eye being the apex of the visual cone, but with the base of the cone itself extended into the upper medium. Vision must then be obscured because the visual rays from the eyes in air cannot pass into another medium without being refracted. On the other hand, were the eye able "without hurting" (*sine laesione*) to be, with its visual "cone," entirely within a continuous fire medium, it would see just as clearly as in air, because there would be no refraction.[9] In other words, the nearer we are to heaven—that is, entirely within the medium closest to it—and lastly in heaven itself, our vision shall be the clearest of all.

Also in section three, Antonino goes on to analyze the physiological anatomy of the eye, seeking similar spiritual comparisons. He knew, for instance, that the *tunicae* ("tunics") or teguments coating the eye including the liquid "humors" inside the eyeball proper, were often thought to number seven.[10] Again like Peter of Limoges, Antonino was quick to grasp the moral significance of this, commenting on how all seven should open sequentially, first to reveal and then to allow the Seven Gifts of the Holy Spirit to enter and pass through the eye, the optic nerve, and then into the spotless soul.

Other similar lessons follow in the same section, and some are rather quaint, but so strong was his desire to milk the science for every possible moral analog that he even applied it to certain popular bigotries regarding disfigurements of human physiognomy. For instance, he observed that persons with "concave" (deep-set) eyes see better, both corporeally and spiritually, than those with bulging eyeballs. The reason, he claimed, is that *species* and rays of light falling on protruding convex eyes tend to disperse and not to focus, so that visual reception in those persons is obscured (*fuscatur*), whereas light rays falling on "concave" eyes are collected more perpendicularly together, so that the eye "is adjusted to its action without hurting" (*sine laesione ad suum actum adjicitur*)—the same principle that explains the efficacy of concave burning mirrors. Stretching this point even further, he observed that, for the same reason that light is diffused from convex surfaces, rain runs off mountains, while water naturally collects in "concave" valleys; the moral message being that mountains signify the vice of pride (*superbia*) while valleys exude the virtue of humility (*humilitas*—which also means "near to the ground")—hence the Virgin Mary properly lowered her eyes in the presence of the Angel Gabriel.[11] As far fetched as this may sound, one should consider that Antonino's words were directed at an audience of Florentine Guelphs, the civic society of merchants and bankers whose prosperous livelihood in their blessed valley metropolis along the Arno River had only recently survived a two-century battle with the insolent Ghibelline robber barons who pillaged with impunity the city's commerce from their arrogant mountain castles in the hilly countryside.

Finally, Antonino turned to the matter of mirrors, about which he had much to say. His long list of spiritual analogies begins, as we might expect, with an analysis of Saint Paul's famous first Epistle to the Corinthians 1:13–12: "For now we see through a glass darkly, but then face to face." It should first be noted, however, that the original Vulgate Latin *per speculum* does not quite mean "through a glass," but more correctly "as in a mirror." The following two words in the Vulgate are *in enigmate*, which, while rendered rather pessimistically in the majestic King James Bible as "darkly," could likewise, as in the post–Vatican II *New American Bible*, be translated "indistinctly."[12] More intriguing, however, is Antonino's own definition of *in enigmate*: "Whereas it is an image, as if a thought or expression is made behind an image; that is, behind a likeness . . . just as they [the prophets] see God or his divine mysteries behind the images and likenesses of sensible things."[13]

Antonino then expanded on *per speculum in enigmate* in seven separate commentaries, each defining a special mirror "by means of which God is seen." Even though Antonino demonstrates that we shall only see God's true reality "face to face" in heaven, in this life we are still permitted to wonder at his extraordinary handiwork, splendid enough if only in mundane reflection:

> And thus we now see God by means of the mirror, as the Apostle saw him when in second Corinthians 3, saying: "While observing the glory of God, that is while we consider God glorious in that mirror, we are transformed from [earthly] clarity into [divine] clarity (*de claritate in claritatem*), that is from one consciousness to another."[14]

There is nothing unusual about the above message per se. It is rooted in medieval Platonism and repeated by Christian preachers and commentators on St. Paul, including Thomas Aquinas, all throughout the Middle Ages.[15] Antonino's reference, however, bears a slightly different nuance. His notion of *in enigmate* as "likeness" is especially subtle because he seems almost to be thinking of mirror reflections as pictures. For instance, he frequently refers elsewhere in the *Summa* to imagining sacred events *in figuram*, "into a figure," the Latin accusative deliberately insinuating a change-of-place transformation from a verbal thought projected in his mind's eye as a picturelike representation.[16] He seems to be implying that pictures, if finally represented with the "clarity" of a mirror reflection, should have the greater potential to "transform" their viewers from worldly to heavenly "consciousness."

This by a medieval theologian who was Archbishop of the most artistically fecund city in all Europe, and who, as prior of San Marco in the early 1440s, had personally watched and probably supervised Fra Angelico as he painted his devotional murals there. Nevertheless, it must be stressed that the "likenesses" Antonino envisaged as mirrored for us by God's grace in this world were hardly realistic in anything like the modern "photographic" sense. I shall henceforth distinguish between Antonino's still medieval spiritual vision and the later, post-Galilean, quite secular concept of "realism."

What effect might Antonino's preaching have had on fifteenth-century Florentines, especially artists, who were also beginning to think of their pictures as mirrors reflecting the grandeur of God's Creation? Is it possible that Lorenzo Ghiberti was so moved by Antonino's optical allegories that he decided to collect and read the optical tracts himself? Was he hoping to apply some of these same *perspectiva* principles in his art just as Antonino did in his sermons? Between 1425 and 1452, Ghiberti operated a busy workshop on a commercial street in Florence that apparently became a meeting place for artists from all over town to come and talk shop, including discussing all the heady new intellectual topics then going around, perhaps even the latest provocative homilies by Fra Antonino.

One painter who surely stopped by was Fra Lippo Lippi (ca. 1406–69), an artist much beholden to the Medici family and who held the Archbishop in high respect. Fra Lippo in fact painted two Annunciations for Lucrezia Tornabuoni, wife of Piero de' Medici. Antonino was her confessor, and it has been argued that he had some influence on their peculiar compositions.[17] Another of Fra Lippi's Annunciations, probably also a Medici commission, ca. 1455, and which now hangs in the National Gallery of Art, London (ill. 19), whether or not it reflected Antonino's influence, clearly owed its iconography to someone with sophisticated knowledge of *perspectiva*, indeed, who might have directed him to Ghiberti, who possessed written instructions.

Close inspection of this work, in any case, reveals that the artist tried to adapt Bacon's complicated *species* theory to illustrate the "scientific" means by which God in heaven dispatched, via the Holy Ghost, his divine seed to the Virgin Mary's womb on earth.[18]

At the very top of this semicircular picture, we notice that the artist has shown the hand of God emerging from a bit of cloud surrounded by a halo-like ellipse rimmed in tiny gilded dots. Below this, as if generated by God's haloed hand, are a series of descending circles likewise rimmed in gilded dots. They seem to overlap one another, exuding thinner streams of gilding from their upper edges as they gravitate to where the dove (the Holy Spirit) is positioned just opposite the Virgin's midsection (ill. 20).

There can be no doubt that Lippi's gilded dots here are his painterly attempt to illustrate the very *species* of God's seed descending from heaven and being "multiplied" through the "medium [of earthly air]" just as Roger Bacon had specified in his well-known theory quoted at the end of the previous chapter.

In the center of the picture, Lippi has his gilded *speciei* encircling the likeness of a dove, the special medium of the Holy Spirit, God's powerful *agens*, now facing the Virgin Mary, not from the conventional upper left as in most Annunciation paintings including others by Lippi himself, but directly op-

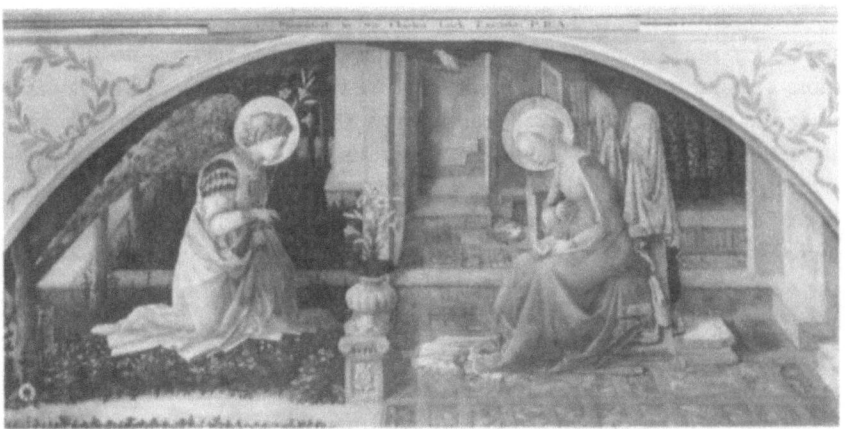

19. Fra Lippo Lippi, *Annunciation*, National Gallery of Art, London, ca. 1455. Photo: Alinari/Art Resource, N.Y.

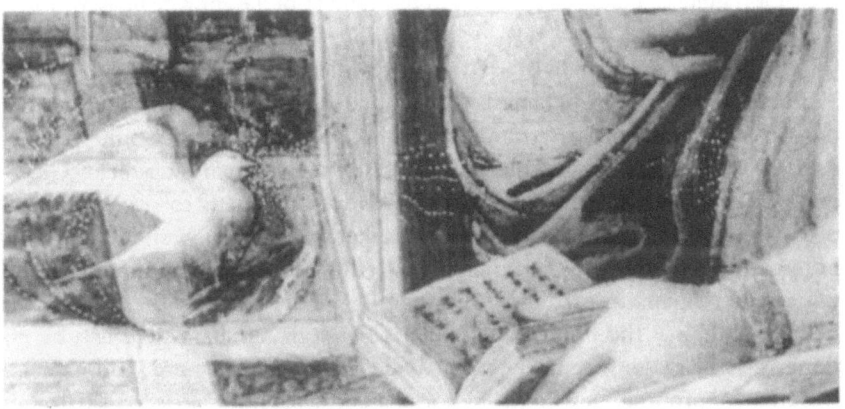

20. Detail of illustration 19, showing "triangles" of gilded dots issuing from the Dove and from the Virgin's womb.

posite her midriff, the position of her sacred womb. Forth from the dove's beak streams another array of gilded motes, this time expanding loosely as a slightly uneven triangle, with a more or less straight line of dots in the center. This is now "refracted" in the direction of the seated Virgin, just as Bacon theorized (even all important "certifying rays" should be bent when passing from a rare to a denser medium, "as is the case of descending from the sky to . . . lower objects."[19])

More remarkably still, from a slit in the Virgin's tunic just below her belt, another triangular spray of gilded dots emerges as if to reciprocate and meet those coming from the dove. From the dove extends its own central ray as if poised to touch the central ray coming from the Virgin. Lippi's imagery here almost exactly replicates Roger Bacon's description of what *species* does between the *agens* and the *patiens* in order to activate the "noble" act of vision. Just as the *species* emitted from the eyes "prepares" the latter for seeing, so, according to Fra Lippo Lippi's unique (as far as we know) application of Bacon's theory, the *species* emitted from the Virgin's womb "prepares" it for the act of divine fecundation:

> Moreover, the *species* of things of the world are not fitted by nature to effect the complete act of vision at once because of its nobleness. Hence these must be aided and excited by the species of the eye, which travels in the locality of the visual pyramid, and changes the medium and ennobles it . . . and so prepares the passage of the *species* itself of the visible object, and moreover ennobles it, so that it is quite similar and analogous to the nobility of the animate body which is the eye.[20]

Furthermore, Lippi depicted the oblique dotted rays coming from the dove in a rather disorderly triangle and actually falling short of the Virgin's dress. This imagery almost literally illustrates a Baconian citation in Ghiberti's *Third Commentary* where he describes what happens to *speciei* that approach the eye obliquely: they are "are broken up (*romperannosi*) . . . and thus not manifest to the eye."[21] Finally, it should be noted that Lippi has the gilded dots emanating from a tiny opening in the Virgin's dress, which is belted just below the bodice and called a *tunica* in Italian. There can be no doubt that he was aware of the optical adaptation of that word, and its neat, Baconian appropriateness to his pictorial analogy of how *species* pass through the ocular outer *tunica* and enter the inner eye through the pupil (*foro*).

CHAPTER FIVE

When Did *Perspectiva naturalis* Become *Perspectiva artificialis?*

A document from the Florentine archives has caught the attention of art historians because of its date, 1413, and of its mention that Filippo Brunelleschi was an "ingenious perspective man" (*prespettivo ingegnoso uomo*).[1] It is actually a letter from a young Tuscan poet named Domenico da Prato telling a friend about his envy that the latter is able to keep company with certain stimulating intellectuals in Florence, while he, Domenico, is bored and isolated in the rustic countryside.[2] Martin Kemp in particular has cited this letter, with its singular word *prespettivo*, as "smoking gun" proof that Brunelleschi painted his famous first linear perspective pictures, the now lost panels respectively of the Florentine Baptistery and the Palazzo Signoria, "on or before" that year.[3] Yet, nowhere in Domenico's document is there allusion to these pictures, or to works of art of any sort.[4]

However, what is mentioned, almost in the same breath by Domenico, and which has not caught the attention of art historians, is that the same friend was simultaneously keeping company with a certain "Antonino," effusively praised by Domenico as "excellent master of sacred theology and profound doctor of the highest wisdom" (*ottimo di sagra teologia maestro e profondo dottore di soma sapienza*). This person, certainly, was none other than Antonino Pierozzi, the future archbishop of Florence and later canonized Saint Antonine, whom we have already met as frequently citing optical principles in his public sermons as metaphor of how God disseminates his divine grace. Antonino would have been just twenty-four years old in 1413, yet Domenico identifies him as already an orator "of marvelous eloquence" (*di maravigliosa eloquenzia*).

Domenico's letter reveals only that in 1413, the thirty-six-year-old Brunelleschi, sculptor, artisan, and already regarded as "a philosopher without books," was engaged with many other Florentine intellectuals like Antonino in the exciting and abstruse discussion of *perspectiva naturalis*. In the following pages, I will argue that in 1413 *prespettivo* could only have referred to optics, and that Brunelleschi was indeed studying the subject diligently at that time, perhaps even under the inspiration of the younger Antonino. And then, some time around 1425, as we shall see, he did apply that science, for the first time at least since classical antiquity, to the art of painting.

Certainly in 1413, *perspectiva* neither in its Latin nor any of its Italianized spellings had any linguistic association with the visual arts.[5] Leon Battista Alberti in his 1435–36 treatise *De Pictura* never mentioned the word even when he explained his own geometric-optical formula for depicting what is seen through his virtual window. If Alberti had thought to name his method at all, he would probably have called it the "cut" or "intersection of the visual pyramid." In fact, the earliest reference to *perspectiva* in association with painting that I have found is in the *First Commentary* of Lorenzo Ghiberti, written around 1452, wherein the author repeats one of the stories from the ancient Pliny's Natural History, Book XXXV, regarding a contest between Apelles and Protogines as to who could paint a more lifelike picture. Although Pliny in the original Latin never mentioned the word,[6] Ghiberti inserted it in his paraphrased Italian version as follows: "Apelles . . . took the brush and composed a conclusion [of the pictorial contest] in perspective appertaining to the art of painting" (*Apelle . . . tolse il pennello e compose una conclusione in prospettiva appartenente all'arte della pittura*).

It is interesting that Ghiberti's qualifying phrase here seems to imply that *prospettiva* still generally meant optics, but that the science as recently applied to painting had once been known to the great artists of antiquity. Nonetheless, he gave no credit for its "rediscovery" to Brunelleschi. Instead, Ghiberti filled his entire *Third Commentary* with indiscriminate glosses lifted verbatim from several medieval *perspectiva* manuscripts then apparently in wide circulation in Florence.

The next time we find our word in elusive reference to painting is in the *Treatise on Architecture* by Antonio Averlino called Filarete (ca. 1400–ca. 1469), originally written in the early 1460s. The Florentine-born author was an avid admirer of Brunelleschi (as Ghiberti was not) and remarked enthusiastically about Fillippo's use of a mirror (*specchio*) in order to compose a picture: "Truly I think it was by this method that Pippo di Ser Brunellesco discovered this perspective, which was not used in other times."[7] What Filarete meant by stressing "this (*questa*) perspective" was that there was now a new "artificial" variant derived from but different from "natural" perspective, especially the optical subscience of catoptrics, which he remembered was Brunelleschi's primary source for understanding how to make pictures in the same way images are formed in mirrors.

Finally, we have Antonio di Tuccio Manetti's biography of Brunelleschi written in the 1480s, twenty years after Filarete's mention and six decades after Brunelleschi's original demonstrations. Although the author would have been too young to have witnessed these demonstrations, he did claim to have held the now-lost panels in his hands. Thus he recorded with confidence that Brunelleschi "propounded and realized what painters today call perspective, since it forms part of that science which, in effect, consists of setting down properly and rationally the reductions and enlargements of near and distant objects as perceived by the eyes of men."[8]

In other words, only by 1480 or so did *perspectiva* (especially in all vernacular spellings) assume its now familiar meaning, and was so defined by Cristoforo Landino, Leonardo da Vinci, Piero della Francesca, and everyone ever since. Curiously, however, even though *perspectiva artificialis* bears the identical family name as *perspectiva naturalis* (or *communis*), the prevailing opinion among many modern scholars is that Brunelleschi did not derive his new construction principally from the optical sciences at all, but rather from his experience as an architect and land surveyor.[9] Moreover, while Filarete and Antonio Manetti both attested that Brunelleschi's first perspective demonstration depended on a mirror, modern scholars have only considered this a clever afterthought, the artist's witty way of showing off his unprecedented "realism" a posteriori.

I will argue instead that not only was the science of *perspectiva communis* crucial to Brunelleschi's experiment but that the mirror, a priori, was the most essential element in its theoretical conception. Furthermore, I also propose that the mirror that Brunelleschi did indeed employ for testing the verisimilitude of his picture was not just to prove that it accorded to the pure geometry of optical catoptrics but that it also reflected the picture *in enigmate*: "a thought or expression . . . made behind an image; that is, behind a likeness," just as Fra Antonino attested was the mirror's ethical function, with all the popular theological implications of Corinthians 1:13: "Thus [the prophets] see God or his divine mysteries behind the images and likenesses of sensible things."

It must be emphasized again that I am rehearsing events that happened in fifteenth-century Florence, still deeply Christian in its thinking, and years before the advent of modern "secular realism." In fact, one might even speculate (no pun intended) that if Antonino himself had witnessed Brunelleschi's

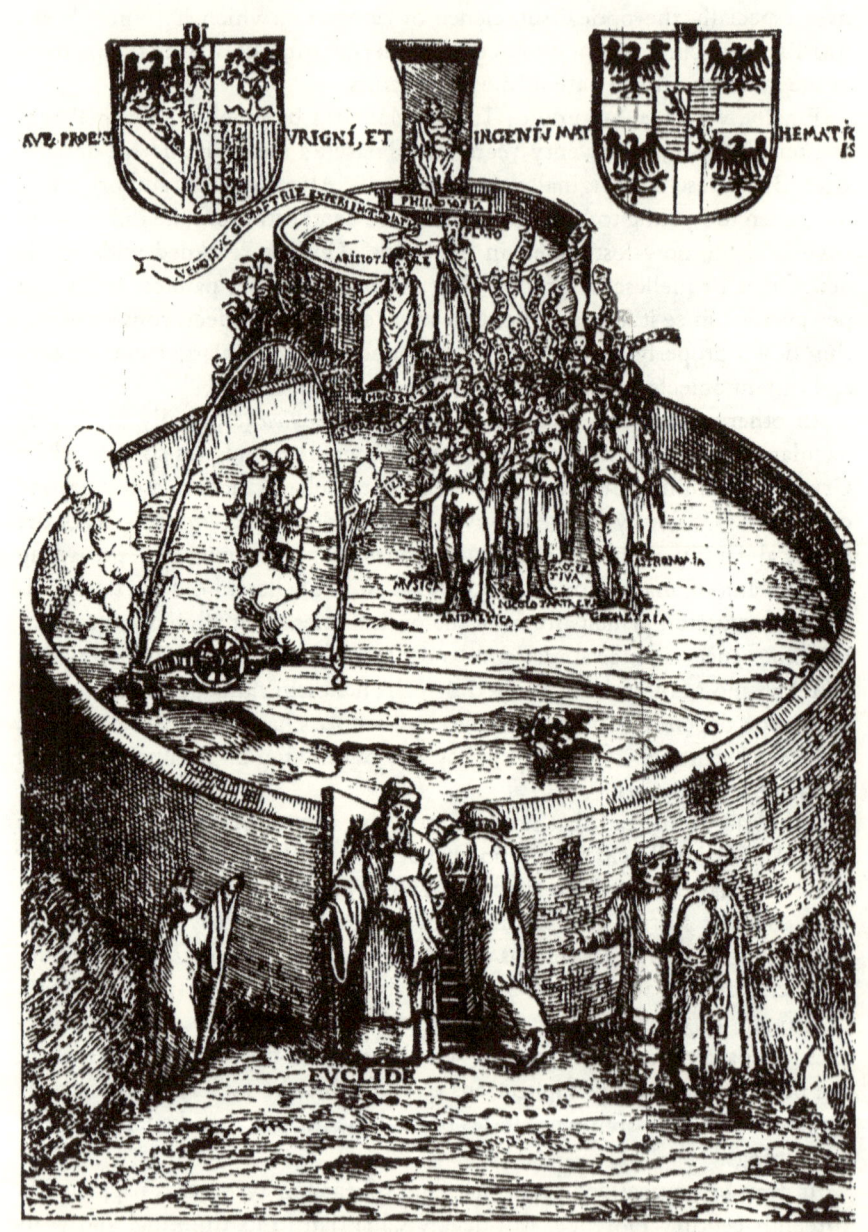

21. Niccolo Tartaglia, frontispiece woodcut, *La Nova Scientia*, Venice, 1537.

first perspective demonstration with its remarkable dependence on a mirror reflection, he might well have considered this as a kind of exegetic "eighth" example after the seven "mirrors" whose spiritual powers he had preached about in his long commentary in the *Summa*. "Whatever can be known about God naturally," he stated, "is manifested in those [mirrors]" (*quod sciri potest de Deo naturaliter, manifestum est in illis*).[10]

It is also noteworthy that by the end of the fifteenth century *perspectiva naturalis*, which by now would also have included *perspectiva artificialis*, was so admired by many European intellectuals that they increased the number of the traditional seven *Artes Liberales* to include the science as the newly honored "eighth." For instance, the Italian engineer Niccolo Tartaglia added a woodcut frontispiece to his 1537 treatise *Nova scientia* (ill. 21), showing himself standing within a walled precinct guarded by Euclid at the entry gate, but ruled by the female personification of Philosophia enthroned at the further end. Tartaglia stands in the center surrounded by ladies in diaphanous dresses who personify the Muses and the Liberal Arts: Arithmetica, Geometria, Musica, and Astronomia at his side, and right behind him (in spite of her name being badly abraded in this poor woodcut) is Prospectiva.

CHAPTER SIX

Brunelleschi's Mirror

It is generally believed that the first of the two lost perspective paintings by Brunelleschi depicted an image of the Florentine Baptistery, then known as the "Church of San Giovanni," showing it as if observed frontally from the opposite portal of the Cathedral of Florence. This octagonal building still stands about 110 feet in front of the Cathedral and directly on an axis with the latter's west façade (ill. 22). As was customary in Italian religious architecture of the Middle Ages, it served as the separate structure where children of Florentine families (usually the most prestigious) should receive their first Christian sacrament. The small piazza between the Baptistery and the Cathedral was (and still is) an important social gathering place, once called the *Paradiso*, or "Paradise," in typological reference to the Garden of Eden where Adam and Eve were created at Genesis. Hence, the whole area was regarded as sacred and identified with birth, original sin, baptism, and heavenly redemption.

The most complete description of Brunelleschi's two perspective paintings is, of course, that by Antonio Manetti, written in the 1480s. It has been frequently translated, and must always be carefully studied anew before any revised analysis of just what Brunelleschi intended and what he accomplished. Here, slightly abridged, is the most accepted version in English of his Baptistery panel account:

> He first demonstrated his system of perspective in a small panel about a half *braccio* [about eleven and a half inches] square. He made a representation of the exterior of San Giovanni of Florence [the Baptistery], encompassing as much of

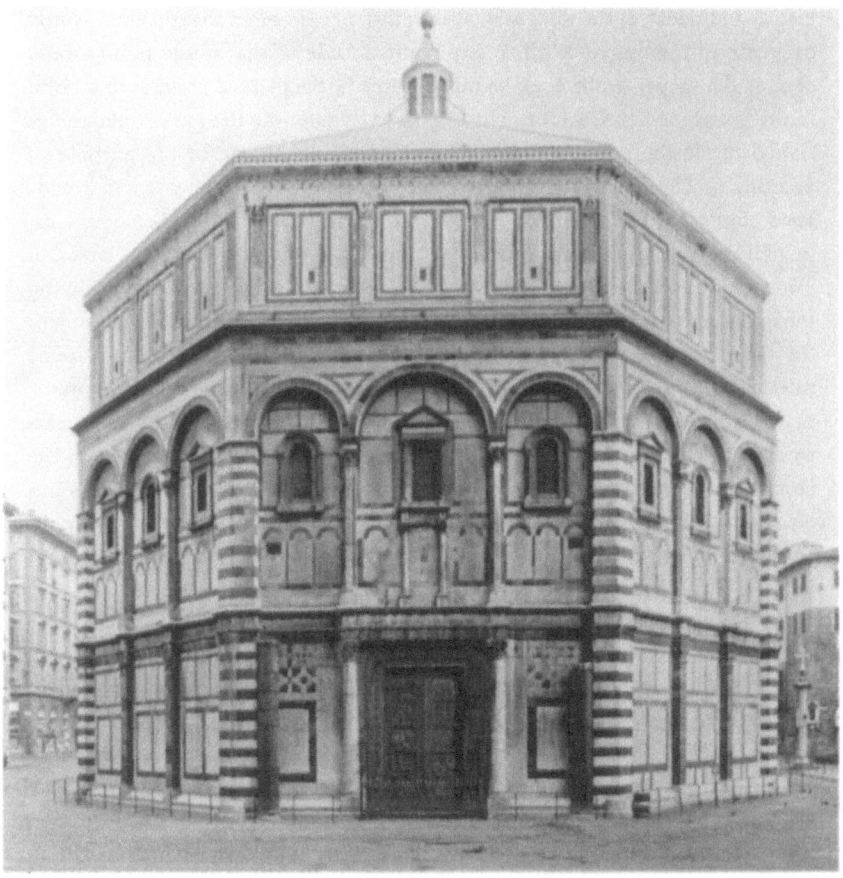

22. The Baptistery of Florence, east façade. Photo: Ralph Lieberman.

that temple as can be seen at a glance from the outside. In order to paint it, it seems that he stationed himself three *braccia* [five feet nine inches] inside the central portal of Santa Maria del Fiore [the Florentine cathedral]. . . . In the foreground he painted that part of the piazza [the Paradiso and the space at either side of the Baptistery] encompassed by the eye. . . . And he placed burnished silver where the sky had to be represented, that is to say, where the buildings of the painting were free in the air, so that the real air and atmosphere were reflected in it, and that the clouds seen in the silver are carried along moved by the wind as it blows.

Since in such a painting it is necessary that the painter postulate beforehand a single place from which his painting must be viewed, taking into account the length and width of the sides as well as in distance, in order that no error would be made in looking at it (since any point outside of that single point would change the shapes to the eye), he made a hole in the painted panel at that point in the temple of San Giovanni which is directly opposite the eye of anyone stationed inside the central portal of Santa Maria del Fiore, for the purpose of painting it. The hole was as tiny as a lentil bean on the painted side and it widened conically like a woman's straw hat to about the circumference of a ducat or a bit more on the reverse side. He required that whoever wanted to look at it place his eye on the reverse side where the hole was large, and while bringing the hole up to his eye with one hand, to hold a flat mirror [specchio piano] with the other hand in such a way that the painting would be reflected in it. The mirror was extended by the other hand a distance that more or less approximated in small braccia the distance in regular braccia from the place he appears to have been when he painted it up to the church of San Giovanni. With the aforementioned elements of the burnished silver, the piazza, the viewpoint, etc., the spectator felt he saw the actual scene when he looked at the painting. I have had it in my hands and seen it many times in my days, so I can testify to it.[1]

Let me rephrase Manetti's vividly remembered detail of Brunelleschi's curious demonstration method. After painting the picture, Brunelleschi drilled a small hole through the panel and then had his viewer look through this hole from the backside of the panel. In his other hand the viewer was to hold a mirror in front of the picture, reflecting the image as seen through the hole. In other words, Brunelleschi did not ask his viewer to admire the verity of the painted picture, but only its mirror reflection. I shall address the reasons for this peculiar presentation in the following pages. But first consider just why Brunelleschi chose this particular building as the subject of his initial demonstration. To be sure there were compelling geometrical reasons, such as the fact that even though it is octagonal in plan, it fits almost perfectly inside a three-dimensional cube, about 110 feet per side including height (up only to the lantern). One could not have found in all Florence a more convenient building of which to make a frontal picture.[2]

Nevertheless, Brunelleschi may have selected the Baptistery because it was also the symbolic device of the *quartiere di San Giovanni*, one of the four municipal precincts into which medieval Florence was once divided (the Baptistery was such a popular public meeting place that it gave its name to the precinct). Perhaps it's not without coincidence that Brunelleschi was chosen as a political representative of San Giovanni in 1425. The flag of his *quartiere*, carried ostentatiously through the city during innumerable civic celebrations, showed the same frontal image of the Baptistery in gold against a blue background.[3] One could almost say that Brunelleschi's perspective panel was intended eventually to be displayed as an updated version of his district's popular *bandiera*.

Furthermore, there may also have been an even more compelling religious motivation, unstated in any of the documents, but surely taken for granted during that still very spiritual age (hence, since everyone at the time would have taken it for granted, there was no need to record it). Since the Church of San Giovanni served as the most prestigious baptismal shrine in the city, its east-facing façade bore special significance because it received the first rays of morning light, reminding Florentines of "the Sun of Justice, Christ our God," as Fra Antonino intoned, "illuminating every man coming into this world" (*illuminantis omnem hominem venientem in hunc mundum*).[4] Moreover, the east doors faced directly onto the main portal of the Cathedral opposite. When both entrances were open, people in the Baptistery could view not only the newly finished west façade recently adorned with life-size sculptures of the Four Evangelists, but the main altar inside the Cathedral itself, while people in the Cathedral could in turn admire the sacred Baptistery, framed by the Cathedral doorway almost like a holy relic in a tabernacle. Processions from the Cathedral on festival days would customarily pass from the latter into the Baptistery through its eastern doors, and out again through its north exit. In sum, as Richard Krautheimer has noted, the east entrance to the Baptistery was the "focus, as it were, for the entire body of sacred buildings on the border between the old and new parts of town."[5]

Indeed the "little hole" that Brunelleschi drilled through the picture, which Manetti so clearly recalled as being situated "at that point in the temple of San Giovanni which is directly opposite the eye of anyone stationed inside the central portal of Santa Maria del Fiore," not only referred to Brunelleschi's careful positioning of his certification point but to its precise centering on the eastern entrance to the Baptistery. If Manetti reported correctly, then there can be no doubt that Brunelleschi did so to identify and connect the exact places where his perpendicular axis had "certified" both the depicted and mirror-reflected Baptistery images, and also to demonstrate to later viewers that he had indeed applied this basic premise of *perspectiva naturalis*. The viewer must be made to believe that his or her eye was optically linked to the same precise "certification" point, just as the artist had focused his eye on the actual Baptistery, according to the fundamental theory of vision as diagrammed in the by-now well-known Euclidian optical pyramid.

Art historians (including myself) have debated whether that hole was intended to indicate what in current linear perspective theory is called the "vanishing point." This is the location opposite the observer's eyes where edges in the visual field perpendicular to the picture plane appear to come together in the distance as in the classic illusion of converging railroad tracks. I still believe that Brunelleschi inadvertently discovered this principle since it is almost a self-evident corollary to the perpendicular axis. I say "inadvertently" because he would not immediately have conceived of it as a purely abstract geometric function. Rather, in the still medieval intellectual ambience of his time, Brunelleschi would have regarded this little hole device as having bestowed on his first perspective picture the same virtue of visual "certification," both *corporaliter* and *spiritualiter*, as the catoptric mirror. Moreover, the little hole actually allowed his viewers to understand how the eye also functioned optically. Through this hole the perpendicular axis passed from the viewer's eye to the Baptistery just as it did through the pupil in the eye itself, on its way to the optic nerve and the *sensus communis*, the "common sense" lobe of the brain where many believed the human soul resided.

There may have been another and perhaps less noble reason why Brunelleschi drilled this little hole, or better, where on the panel he chose to drill it. Almost at the same moment that he was preparing this demonstration, the Florentine Wool Merchants' Guild, the Arte di Calimala that had charge of maintaining the Baptistery, was installing the building's new doors that Lorenzo Ghiberti had just sculpted and cast in bronze for the eastern entrance. Some twenty-three years before, Ghiberti had beat out Brunelleschi in a close competition for that then very prestigious commission that had subsequently enormously elevated Ghiberti's social importance. No doubt Filippo must have felt a tinge of jealousy as his rival was basking now in such triumph (in fact having been granted without open competition a lucrative commission for yet another set of Baptistery doors, the famous "Gates of Paradise," the contract for which was signed on January 2, 1424).[6] How indeed this must have fueled the mutual antagonism that was to last throughout both their long careers. One may even conjecture that Brunelleschi chose to drill the small hole through the back of his panel so that it cut right through the image of Ghiberti's east doors depicted on the painted side, thus compelling the viewer "to see through" (*perspicere*) his rival's handiwork, and view Brunelleschi's own eye, rather than the door, in the mirror reflection. Brunelleschi in effect was diminishing Ghiberti's achievement. By showing off his newfound ability to conceive the form of the whole building in sacred catoptrical geometry, he had declared himself intellectually superior to the mere craftsman carver of decorated doors.

Ghiberti, on the other hand, would not have allowed this slight to go un-answered. In his next set of Baptistery doors, the famous "Golden Gates of Paradise" completed some twenty-five years after his first success, he did take account of Brunelleschi's new perspective projection method in several of his relief panels, but always with a certain idiosyncratic interpretation as if he himself knew more about how optics should be applied to art than Brunelleschi.[7] To Ghiberti's delight, his second pair of doors was judged so beautiful that the Wool Merchants' Guild decided to place them instead in the iconic eastern portal, and move his earlier set to the northern entrance. About halfway up the inner edges of the door frames he then cast to enclose his new set of panels, Ghiberti carved miniature portrait heads of himself and his son, Vittorio, fully in the round peeping rather provocatively out of a pair of illusionistic peepholes (ill. 23), as if to mock Brunelleschi's similar peep-hole through the same portal depicted in the first perspective panel.

Nonetheless, Brunelleschi's demonstration permitted viewers to believe that they had penetrated the very "enigma" of the mirror, to see both the virtual reflection and actual Baptistery "face to face" behind the reflection, just as Saint Paul had preached. His small handheld panel of the Baptistery astonished fifteenth-century Florentines because it revealed not just a supe-rior likeness but rather because the artist's perspective seemed to enhance as

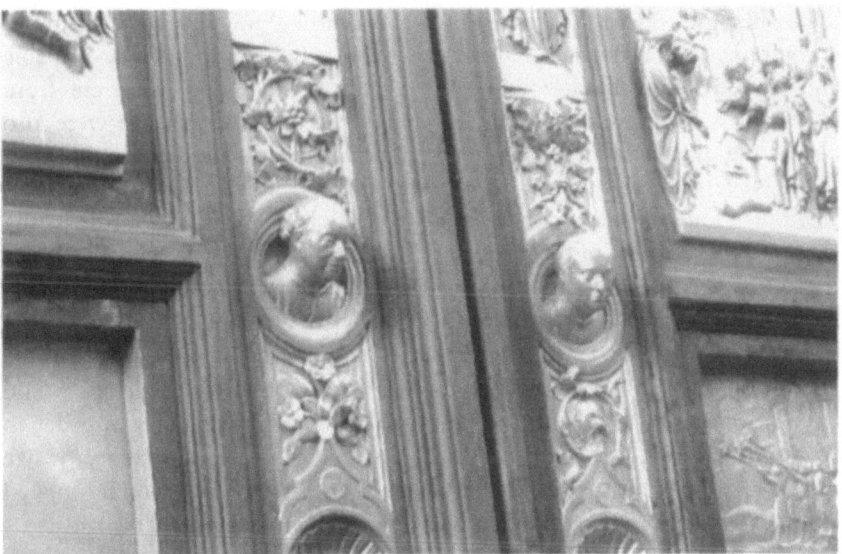

23. Detail, self portraits of Lorenzo and Vittorio Ghiberti, *Gates of Paradise*, Baptistery, Florence, ca. 1452. Photo: author's collection.

never before the symbolic importance of geometric measurement. As Antonino put it, Spiritual geometry works to measure temporal things . . . It
measures dimensions not as quantities but as virtues within God" *(Mensurare
temporalia fecit geometria spiritualis . . . Dimensiones non quantitativas, sed virtualis
mensurat in Deo).*[8]

By focusing his perpendicular axis, across the typological *Paradiso*, and
onto the eastern entrance to the Baptistery with its own sacramental symbolism, Brunelleschi was identifying his certification point as if it were in the
very eye of God as he created the first human beings during the sixth day of
Genesis. Brunelleschi's viewers were enticed to believe themselves envision-
ing the very process by which "the prophets see God or his divine mysteries
behind the images and likenesses of sensible things."

Finally, a word should be said about the small size of Brunelleschi's first
perspective panel and its relation to the mirror, and of the special effect of
looking through the hole in the back of the panel at the mirror reflection of
the Baptistery picture. According to Manetti, this panel was only a half *braccio* square; that is, a bit less than twelve inches on a side, while his second
panel, of the Palazzo Vecchio, was nearly twice as wide. No doubt Brunelleschi
chose the lesser size for his first demonstration because that picture needed to
be close to the size of the mirror, which fifteenth-century technology could
not yet manufacture much larger.

To be sure, during the early fifteenth century, mirrors were hardly as
commonplace as today. Crystal mirrors were only beginning to be available
and tended to be small, round, and convex because they were cut from
blown glass balloons. However, flat metal mirrors of larger size were also
obtainable although they were more cumbersome and less luminous than
glass. Nevertheless, I would hypothesize that a one-half *braccio* square polished metal mirror was exactly what Brunelleschi used, at least for deducing
the optical geometry of the Baptistery's reflection.[9] Still, the reflective quality of any mirror at the time, either glass or metal, was poor by modern stan-
dards. Surface imperfections and distortions were unavoidable and must have
rendered any reflected image somewhat "enigmatic" to say the least. One
could presume, however, that the famously resourceful Brunelleschi would
have spared nothing to find the best mirror in size and quality in order to
work out his method. In any case, such a metal mirror of one-half *braccio* size
would have been rather heavy, and he could only have worked on it propped
on an easel. This could mean that he needed yet another mirror for the demonstration, light enough to be held in one's outstretched hand, in which
event the second mirror would more likely have been made of glass. Although Manetti in his biography of Brunelleschi stated that this latter view-

ing mirror was "flat" (*piano*) one must understand that he was speaking of mirrors relative to his own late fifteenth-century time when flat glass mirrors were far more common. Earlier in the century, however, the most available were circular in shape and convex like the household mirrors often depicted in fifteenth-century paintings, as in Jan Van Eyck's *Arnolfini Wedding*, for example (ill. 24).

If Brunelleschi's half-*braccio* square painted panel is understood as the in-

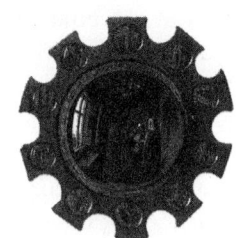

tersection of the artist's visual pyra-mid as he faced the Baptistery from the portal, then, in order to achieve the same optical effect as seeing the object itself, he (and his first viewers) should have been able to hold the picture before the eyes exactly one-half braccio away; that is, in proper proportion to the actual dis-tance between themselves standing in the Cathedral portal and the ac-tual Baptistery façade—a "small *brac-cio*" signifying a "regular *braccio*," as Manetti quaintly expressed the then novel notion of architectural scale. But Brunelleschi wanted not the pic-ture but the mirror, whatever its form, to be held before the viewing eye. The picture instead should be turned around and held with the other hand against the viewer's face so that he or she might peek through the unique hole in the back and ob-serve the painting on the obverse side only by its reflection. If Brunelleschi performed this demon-stration by having his viewer stand in the very spot in the Cathedral doorway from where the picture was painted in the first place, then, when the mirror was removed, the viewer should see the real Baptistery through the hole, and supposedly marvel at

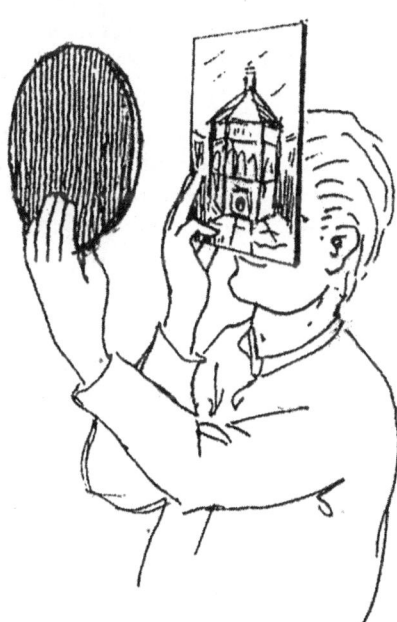

24. Brunelleschi viewing his first perspec-tive panel in a convex mirror. Drawing and montage by author.

how closely it compared to the mirror reflection.[10] Considering the likely distortions in Brunelleschi's demonstration mirror, especially if it were made of blown glass and convex, we may wonder why he chose this peculiar manner of presentation to "prove" the "realism" of his painting. How could his experiment have been judged as being so successful—that is, if we consider it only from our modern expectation of what constitutes true scientific "perspective"?

To be true to historical "perspective," on the other hand, we must judge Brunelleschi's demonstration not from what modern eyes would want to see, but rather from the expectancy of a fifteenth-century Florentine who had never before seen an optically corrected perspective picture. The increasingly popular standard for measuring pictorial "realism" in that transitional medieval/Renaissance century, I submit, was not by direct comparison to the phenomenal world, but by comparison to the novelty of mirror reflection. In the transalpine north this is even more evident. Not only Jan van Eyck (ca. 1395–1441) but Robert Campin (ca. 1378–1444) and Petrus Christus (d. 1472/3), as well as other artists in the region sought to achieve with oil glazes the same surface sheen and detail of mirrorlike reflection (but not catoptrical geometry, at least until after the 1430s[11]), and sometimes included images of reflecting mirrors in their paintings. There are even depictions of artists painting their own self-portraits from convex mirror reflections. One of the earliest is from a French manuscript dated before 1404 showing "Marcia," a famous ancient Roman woman painter described by Pliny.[12]

In sum, Brunelleschi's test for the success of his demonstration probably depended on his viewer comparing the reflection of his panel not to the Baptistery itself but to the Baptistery's own reflection when the same mirror was turned around to face the building.

There was still another problem that may well have been raised by the more mathematically minded observers of Brunelleschi's demonstration. How far should the mirror be held before the picture in order that the latter's reflection exactly replicates the building itself? If the mirror was simply an inert surface, the answer would be easy: a half *braccio* before the eye, the same as if it were the picture. But the mirror does a peculiar thing with whatever it reflects. As far from the mirror as the object being reflected is before it, the reflection will appear exactly that same distance within the mirror's virtual space (ill. 25). Therefore, the image of Brunelleschi's painting would not appear directly on the mirror surface, but as if it were behind the mirror surface, by exactly one-half *braccio*, and thus a full *braccio* before the observer's eye.

Many modern scholars have likewise puzzled over this conundrum,
because the only way Brunelleschi
could have solved it would have been
to hold the mirror just a quarter *braccio*, less than six inches, in front of
the picture against the viewer's eye.
Obviously, this would have made it
very difficult to see the picture's reflection at all, and so we are left to
conclude that Brunelleschi either ignored the optically correct distance
requirement or he accepted the problem as actually serving his principal
reason for having the mirror in the
first place.

If I am right that Brunelleschi was
under the influence of Fra Antonino's moralized optics at this time,

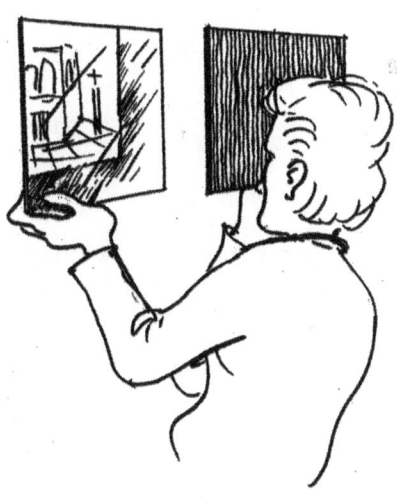

25. Brunelleschi viewing his first perspective picture in a mirror, holding it a half
braccio before his eye. Drawing by author.

what better proof might he have had of Saint Paul's Epistle to the Corinthians? One may even imagine Brunelleschi, as he led his viewers through the
optical technicalities of his first perspective demonstration, offering this les-
son to anyone who might have noticed this or any other of the possible reflection discrepancies: "In our mortal world, just as in my mirror, you see the
Baptistery only enigmatically. Not until you are in heaven face to face with
God, will you at last behold its true reality."

CHAPTER SEVEN

Brunelleschi's Method

Of the many attempts to recapitulate how Brunelleschi painted his first perspective picture, the hypothesis most frequently cited is that he constructed it by means of what is today called "orthographic projection," a method architects employ in order to make three-dimensional views of their buildings (ill. 26). One simply aligns on the drawing board a scale plan next to a same-scale elevation of the building in such a way that a third view can be projected by intersecting the coordinates between them. The result will indeed appear three dimensional and be in perfect geometric perspective.

Such a solution, as here illustrated, was first proposed by Richard Krautheimer many years ago, and has recently been championed again by David Summers.[1] The reasoning behind this reconstruction is based on two facts. First, Brunelleschi was an architect of great skill, and, during the early years of the fifteenth century, according to Manetti, he went to Rome with his friend, the sculptor Donatello (ca. 1386–1466), where they surveyed classical buildings, measured the plans, calculated the heights, and drew the details by means of "strips of parchment which were laid over the charts [to divide them] into squares with numbers and letters that only Filippo understood."[2]

The second reason is based on Giorgio Vasari (1511–74) who included a biography of Brunelleschi in his famous *Lives of the Most Eminent Painters, Sculptors, and Architects*, written in 1550 and revised in 1568. Vasari, who lived a full century after Brunelleschi's death and deduced his material solely from secondhand sources, described the experiment as follows:

54

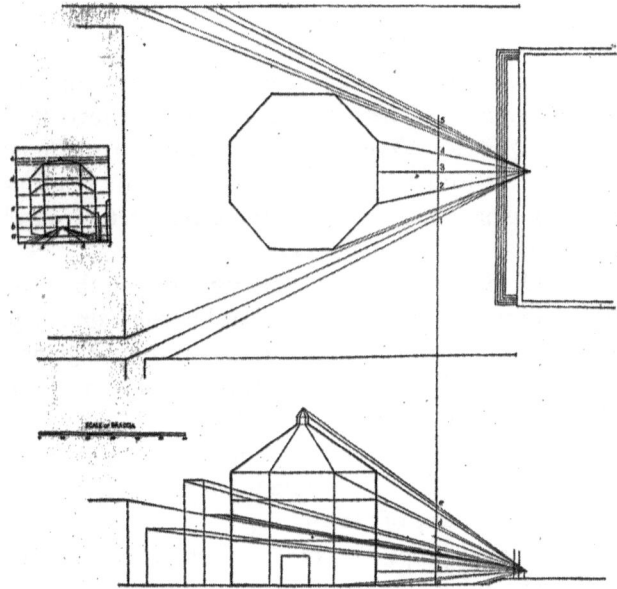

26. Hypothetical reconstruction of Brunelleschi's first perspective panel by Richard Krautheimer; from R. Krautheimer and T. Krautheimer-Hess, *Lorenzo Ghiberti*, 1970. Reprinted with permission from Princeton University Press.

[He] gave considerable attention to the study of perspective, the rules of which were then very imperfectly understood, and often falsely interpreted; and in this he expended much time, until at length he discovered a perfectly correct method, that of taking the ground plan and sections by means of intersecting lines, a truly ingenious thing, and of great utility to the arts of design.[3]

The problem with such a reconstruction has already been addressed. If Brunelleschi derived his new projection method from drawing techniques learned in architectural practice, why did all his near contemporaries call it *prospettiva*? Why not *scenographia*, which many antiquarians of the time also knew about, and would seem to have better defined such an architectural procedure? Obviously, it must have been assumed from the beginning that this new idea related directly to the science of optics. Furthermore, Vasari's account should be taken with a grain of salt (as with many of his anecdotes in the other biographies). In fact, he never saw Brunelleschi's two panels (by 1550 they had already been lost), and, as noted above, his knowledge of them followed largely from the same fifteenth-century accounts that we have already examined, none of which say anything about such a projection method. Also, Vasari makes no mention of the mirror or the hole in the Baptistery picture, which according to Manetti were the unique features of Brunelleschi's first demonstration. In fact, Vasari's account assumes that Brunelleschi had already understood the basic optical principles of horizon line and certification point, and that whatever architectural projection method Vasari adduced to him must have been applied afterward.

Moreover, building plans drawn to scale were rare in the early fifteenth century.[4] Whatever techniques Brunelleschi had devised for measuring ruins in Rome seem to have been more akin to grid-plan stratigraphic drawing such as archaeologists employ today for plotting the reconstruction of ancient sites. Indeed, Brunelleschi might even have devised a mapping system of his own, based on the Ptolemaic coordinate method of cartographic-grid projection, another antique science heretofore unknown in Western Europe but just recently retrieved by way of Florence.

In 1400, a Greek manuscript copy of the *Geographia*, originally composed by Claudius Ptolemaeus (fl. 127–145 AD), inadvertently arrived in the city from Constantinople, which described by application of the same optical principle of certification how to project the curving surface of the spherical earth upon a flat plane, and then to draw maps that divided the various regions of the earth into grids composed of vertical meridians and horizontal parallels, making it possible to position every individual location by their paired coordinates on this grid (ill. 27). Florentine intellectuals, surely including Brunelleschi, were just as excited about this new cartographic technology as they were about *perspectiva*.[5]

More than forty years ago, in preparation for my eventual 1975 book, *The Renaissance Rediscovery of Linear Perspective*, I, with the help of colleagues, attempted to recapitulate Brunelleschi's original perspective experiment as Manetti remembered by actually positioning a one-half *braccio* square mirror (about eleven and one-half inches on a side, the same size as the first perspective panel described by Manetti) some three *braccia* (about five feet nine inches) within the portal of the Florentine Cathedral facing the Baptistery, and photographing the reflection from a height of about three *braccia* and at a distance of one-half *braccio* (ill. 28).[6]

Unfortunately, our old model Nikon camera could at that time only take rectangular photographs, so the upper part of the third story of the more or less square frontal view of the building had to be added. Nor could the camera compensate for vertical convergence distortion of the building's sides. Nonetheless, our photo did reproduce an image that must have been close

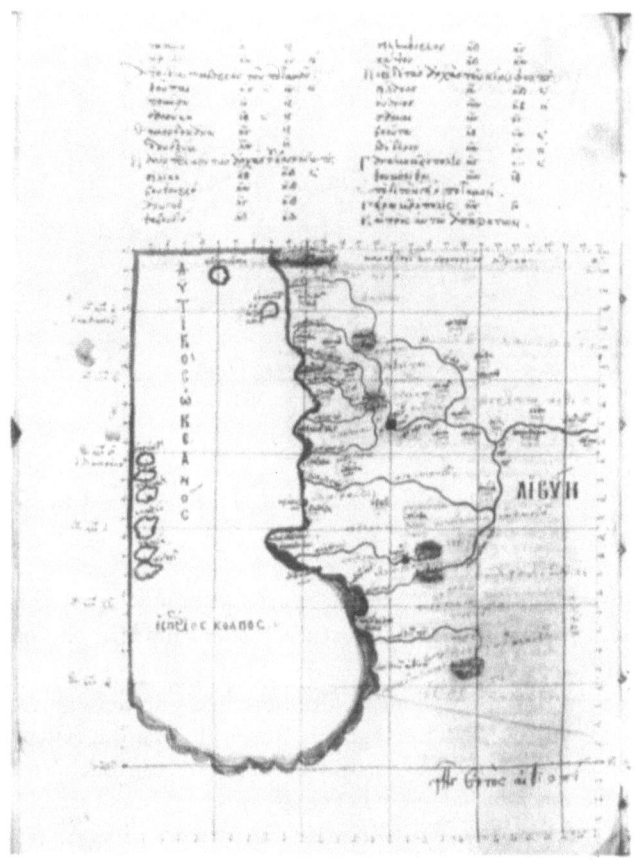

27. Map page from the *Geographia* of Claudius Ptolemaeus
(Ptolemy), Florence. Biblioteca Medicea Laurenziana, ms. Plut.
28.49, C.11, fol. 51v. By permission of the Ministero per i Beni e le
Attività Culturali.

to what Brunelleschi saw in his own mirror when he set up his original
demonstration. I also enhanced my photograph by adding an overlay drawing, to indicate by a white dot the "certifying" focal point of the reflected
camera (and exactly where Brunelleschi would have drilled that famous
hole), and also a line passing through this point defining what is today called
the sensible horizon (and what Alberti later called the "centric line"), at the
same eye level. Brunelleschi's apparent discovery of this coincidence was his
unique contribution to the science of optics, since no scientist or artist had
ever observed the connection before. What Brunelleschi now could realize

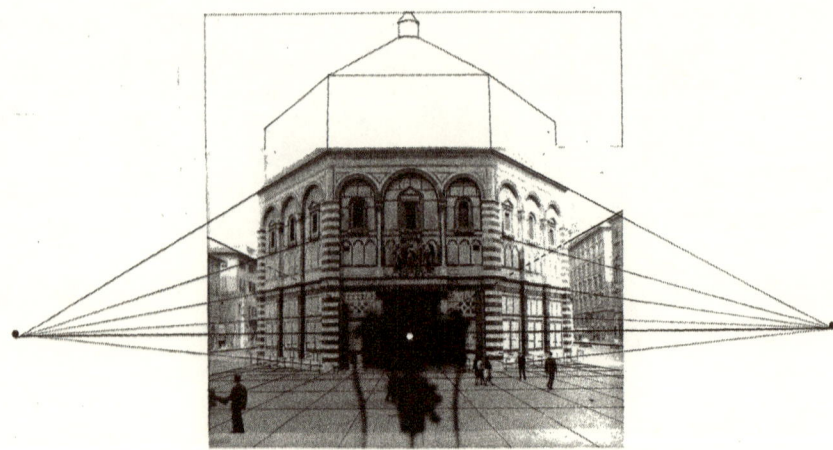

28. Reconstruction of Brunelleschi's first perspective picture. Photo and overlay by author.

for the first time in the history of world art was how to apply this principle to the construction of an artificial picture, just as a modern photograph does automatically.

My photo overlay also includes a foreshortened grid drawn over the foreground piazza with its own orthogonals converging on the camera lens, and also the points on the same horizon line to which the top and bottom edges of the two visible forty-five-degree diagonally directed lateral faces of the Baptistery appear to converge, revealing that the original panel must have demonstrated the rudiments of what is today termed "oblique" as well as "frontal" perspective. As a result of this experiment, I am convinced that Brunelleschi took advantage of a traditional artistic convention, a squared pavement as ground plane under both his images of the Baptistery and the Palazzo Vecchio perspectives, even though the actual piazzas surrounding these edifices at the time were probably not paved with evenly squared stones.[7]

Nevertheless, such depictions of squared ground planes and ceilings were already commonplace in late medieval European art. Even when optically incorrect, they had long since been employed by artists as appropriate settings on which to stage sacred subject matter. Furthermore, the Sienese painters Piero (active 1306?–1345) and Ambrogio Lorenzetti (active 1319–1347) were able to render these empirical conventions so naturally that they could almost be mistaken for true perspective, for instance the intricately tiled pavement in Ambrogio's *Presentation in the Temple* dated 1342, now in the Uffizi Museum, Florence (ill. 29).

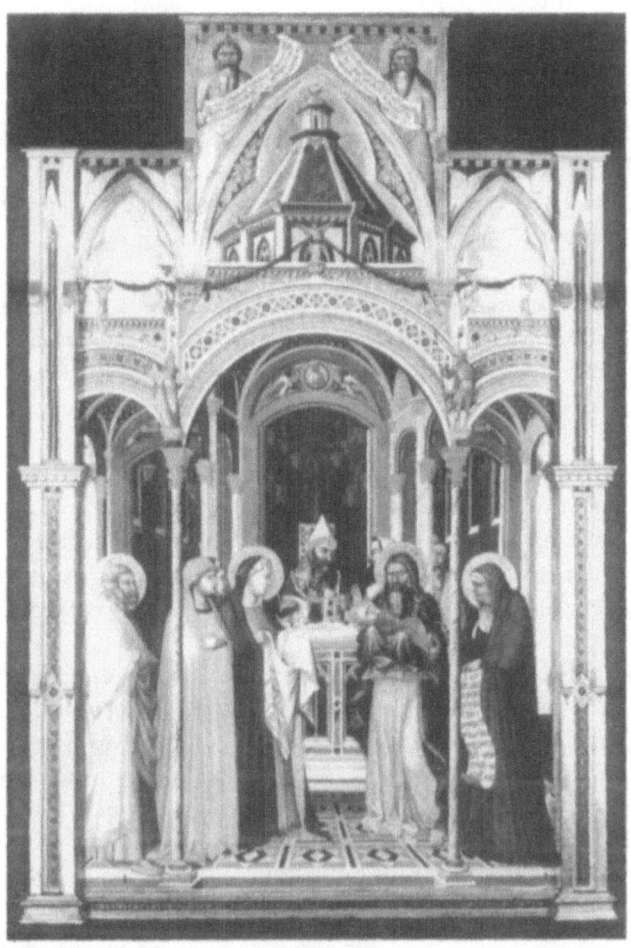

29. Ambrogio Lorenzetti, *Presentation in the Temple*, Galleria degli Uffizi, Florence, 1342. Photo: Gabinetto Fotografico del Polo Museale Fiorentino.

Brunelleschi's inclusion of this old convention in both his panels, whether or not picturing the true condition of the actual piazzas, would have lent his images a certain iconic cachet, and therefore make them more "believable" to tradition-bound Florentine viewers.[8] Even had he wished to depict in his two panels the piazzas as they looked at the time, he could simply have painted over his underlying construction drawing. But more important is the fact that the easiest way to draw the perpendicular portions of a building was

by simply projecting them upward from a foreshortened plan already out-
lined on a the trapezoidal grid, as Leon Battista Alberti was to explain again
in his 1435/6 treatise on painting.

In any event, Brunelleschi still had to perform one more operation in order for his picture to imitate exactly the mirror image. This was what today
is called the "distance point construction," a concept that was never specifically addressed in the traditional literature of medieval *perspectiva naturalis*.
Nevertheless, as Manetti recalled:

> Since in such a painting it is necessary that the painter postulate beforehand a
> single place which his painting must be viewed, taking into account the length
> and width of the sides as well as in distance, in order that no error would be
> made in looking at it (since any point outside of that single point would change
> the shapes to the eye).[9]

In other words, what Brunelleschi had also to discern from his mirror image was how and why the distance between viewer and mirror surface always
controls the fictive tilt of the reflected ground plane in the mirror's virtual
depth, steep if viewed close-up, and more leveled if viewed from further
away.

I submit that there is compelling evidence of Brunelleschi's original "distance point" solution in the very half-*braccio* size he chose for his first demonstration panel. It would have easily occurred to him if we assume that, before
beginning the painting, he had carefully measured the whole Baptistery piazza, just as he had ancient Roman sites many years before with his friend
Donatello, perhaps resulting in a drawing such as that of the same piazza by
Bernardo Sansone Sgrilli in 1733 (ill. 30).

Had Brunelleschi then drawn a simple visual triangle encompassing the
frontal view he wished to paint, as I have overlaid on the Sgrilli plan, he
would have immediately observed that the overall width of the visible façade
of the Baptistery, straight in front and slanting to either side at forty-five degrees, almost exactly equaled the Baptistery's distance from the portal of the
Cathedral opposite. This was certainly another reason why he chose the Baptistery for his first perspective representation. It permitted him a comfortable
1:1 ratio between the width of the Baptistery and his distance from it, and the
size of his mirror and his distance in front of it as he copied the reflection
onto his painting.

As mentioned earlier, the illusion of a floor or even a ceiling plane as a
gridded checkerboard-like surface was a well-recognized artistic convention
long before Brunelleschi. Many early painters sensed empirically the conver-

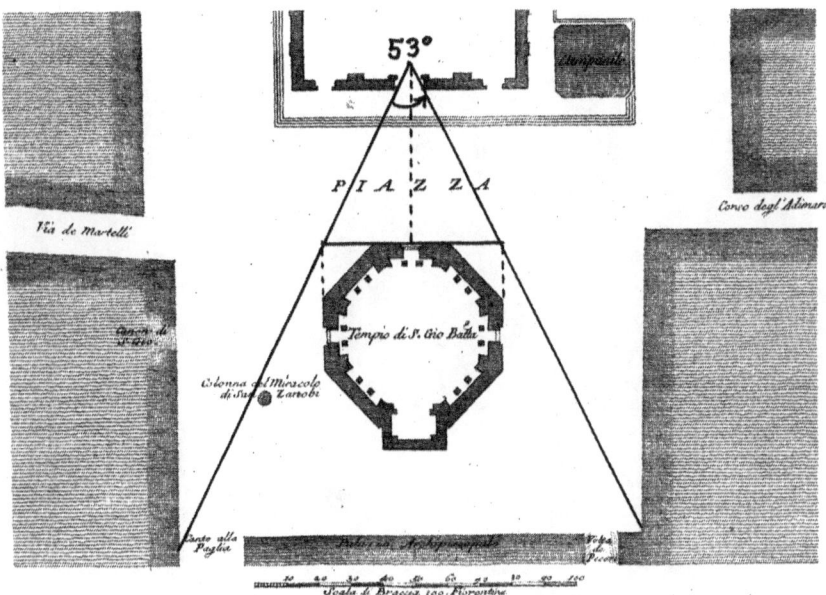

30. Site plan of the Cathedral Piazza, from Bernardo Sansone Sgrilli, *Descrizione e studi dell'insegne fabbriche di Santa Maria del Fiore . . .* , Florence. Reconstruction by author of Brunelleschi's visual angle looking from the Duomo toward the Baptistery.=

gence effect, even realizing that the transverse spaces between parallel edges seem to diminish in a regular manner the further away they are in distance, and their receding diagonal edges come together in clusters if not in single points. During the fourteenth century several empirical workshop systems were devised by artists for depicting the illusion of a squared floor or ceiling receding into the virtual depth of the picture. Leon Battista Alberti in the final version of his treatise on painting mentioned one of these, which he severely denounced as optically incorrect. It was called *subsesquialtera*, an ancient arithmetical method for expressing the expanding ratios between certain numbers that some painters apparently were applying arbitrarily in order to calculate the narrowing spaces between the receding transverse parallels in their depictions of gridded surfaces.[10]

Another such empirical device has come to be known as the "bifocal construction." It was detected and described many years ago by Italian art historian Decio Gioseffi and French scholar Robert Klein, particularly in a fourteenth century fresco in the Franciscan Basilica at Assisi.[11]

On the slightly curved vaulted wall of the north transept in the lower church an anonymous artist, certainly in the Lorenzetti circle, set out a "cat's

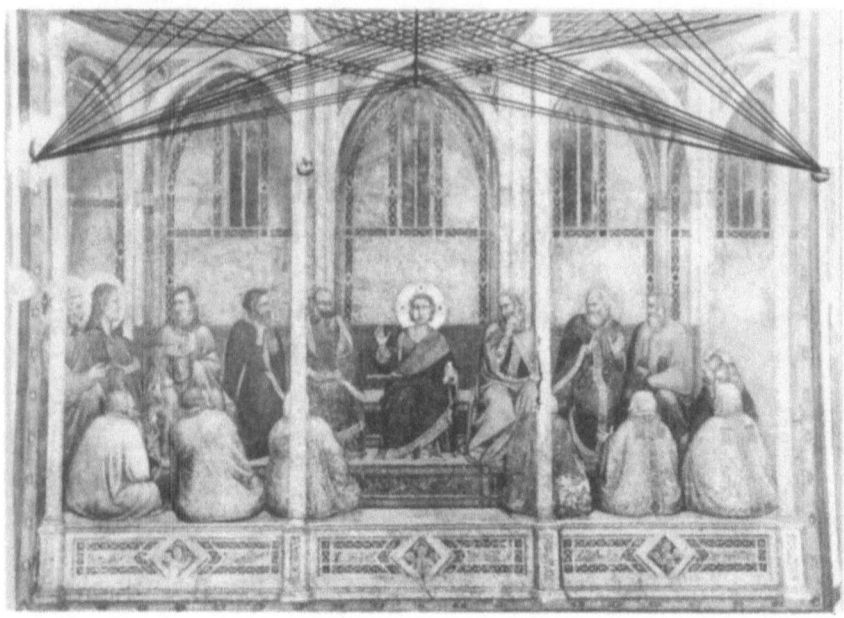

31. Reconstruction by author of the bifocal "perspective" projection: anonymous Sienese master, *Jesus among the Doctors*, lower church, Assisi, early fourteenth century.

cradle"-like network of inscribed lines that I have redrawn as an overlay in his depiction above of *Christ among the Doctors* (ill. 31). There are still two small iron rings affixed in the wall, one each on either side of the painting, to which the artist tied his snap strings for ruling in his gridded ceiling.

Another clear example of this convention is to be traced in a panel by the Umbrian artist Ottaviano Nelli (1375–1444?) depicting the *Circumcision of Baby Jesus* painted sometime between 1415 and 1425, and now in the Vatican Art Gallery (ill. 32). Either the artist in his faraway hometown of Gubbio was remarkably prescient, or he had already heard of Brunelleschi's experiments. Indeed, he apparently did realize how perspective convergence could heighten the picture's drama by focusing the viewer's gaze directly on the sacred surgery.

In any case, all the artist need do to set up this technique was to fix two pins, one each at the same height on either side of the surface to be painted, and then divide with more pins the baseline of the same surface into a select number of equal segments. From both pins at the sides, the painter would then connect snap strings to each of the pins on the baseline. Wherever the strings crossed on the surface to be painted, the artist would draw lines both

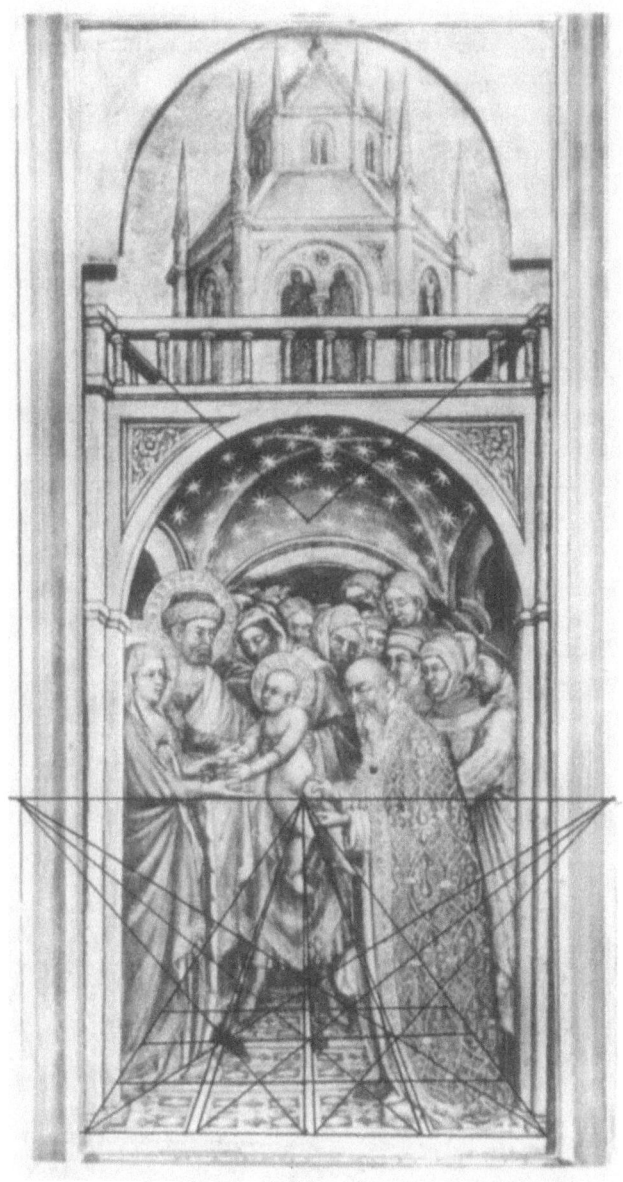

32. Reconstruction by author of the bifocal "perspective" projection: Ottaviano Nelli, *Circumcision of Baby Jesus*, Pinacoteca, Vatican
City, ca. 1415/25.

horizontally and diagonally through these intersections. This automatically forms a gridded trapezoid, the diagonal edges of which converge on a point in the center on the same level as the pins at the sides of the picture surface. Nevertheless, no painters before Brunelleschi, not even Nelli or the precocious Lorenzetti brothers, had any idea as to the optical reasons why such a convention looked so realistic.

Let us now presume that Brunelleschi, standing within the Cathedral doorway, faced his mirror toward the Baptistery and then viewed the reflection from the same one-half *braccio* distance as my camera in illustration 28. If next he traced directly on the mirror surface the reflections of the forty-five-degree slanting edges of the two facing oblique sides of the building and followed them to where they seemed to be converging top and bottom, he would have noticed that they came together in separate points, left and right, at the same precise level as his centric certification point. As mentioned, this was a crucial observation that provided Brunelleschi the essential key to how to replicate precisely a mirror reflection in a picture. Furthermore, he must have noticed that these two points were spaced just about one-half *braccio* on either side of the centric certification point. Indeed, he must have had to add extra drawing surface at the sides of his mirror to record these extensions. What was he then to make of this curious latter coincidence, that the lateral points where each of these oblique edges of the Baptistery appeared to converge were not only at the same level as his centric viewpoint but at the same distance from it as his own distance from the mirror?

Being an experienced surveyor, Brunelleschi was surely familiar with a common practice of that time for determining the breadth of a river or dis-

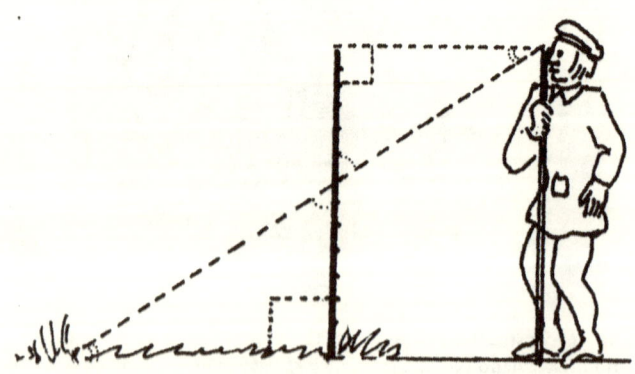

33. Surveyor measuring with a staff. Drawing by author.

tant piece of land by sighting its far and near boundaries across a graduated staff, similar to what today is called a "stadia rod," and triangulating these measurements in proportion, as in my illustration 33. Once he realized that parallel oblique edges appear in mirror reflection to converge at the same horizon level as the certification point, another geometrical coincidence should quickly have become apparent. If, as I have hypothesized, he decided beforehand to lay out his painting on a conventional gridded ground plane, and had already drawn diagonal lines from evenly spaced divisions of his panel's baseline to a centric certification point, he would surely have noticed that the centric vertical line through this certification point functioned just like a staff being employed by a surveyor standing at the same eye level a half *braccio* on either side of such a staff. As the "visual rays" from each of the evenly spaced divisions of such a ground plane crossed the staff on their way to the surveyor's eye, he would mark that place and note that the distances between his markings grew gradually shorter the further away the spaced division. If he diagrammed the visual rays as they crossed each other diagonally from both sides of the staff, and then drew horizontal lines across the diagonals, he would have a drawing that looks quite like my illustration 34, which displays what I'm trying to say far better than words.

What Brunelleschi had suddenly solved was the puzzle of the old bifocal method. Although a centric vertical in that empirical technique might likewise be thought of as a surveyor's staff, the lateral points, however, were usually placed at either edge of the picture frame, therefore only half the picture's width away from its centric line. This implies a very short distance between artist and image, which is what often causes the illusion of squared ground planes in these early Renaissance paintings to look so steep.

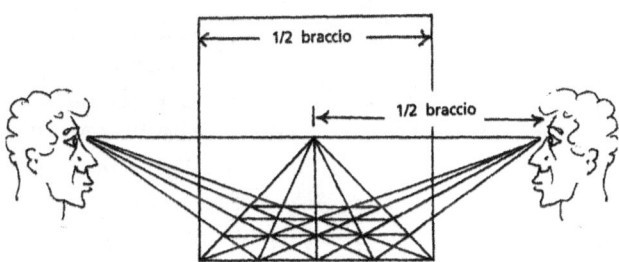

34. Brunelleschi solving the bifocal construction. Drawing by author.

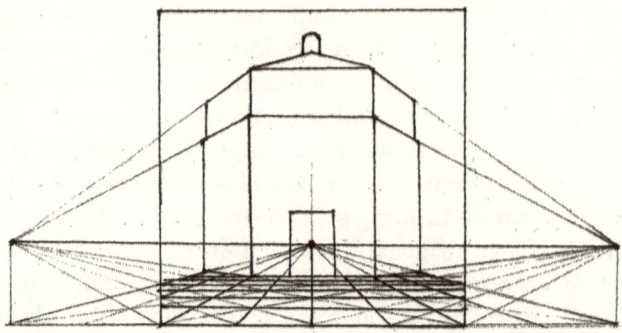

35. Brunelleschi's first perspective layout. Reconstruction by author.

Nevertheless, Brunelleschi would now have discerned that this same structure actually underlay his mirror image of the Baptistery, which needed only to be readjusted to his true eye level and viewing distance (ill. 35). Once he applied the same "cat's cradle" geometry to the wider space between the two "bifocal points" of the Baptistery reflection, and marked their crossing points on the edges of his panel, he could then proceed to copy the rest of the building's visible details directly from the mirror reflection, confident that their collective "perspective" conformed to the actual catoptrics of the mirror as viewed from his height at half-*braccio* distance (ill. 36).

Finally, let us probe an important piece of evidence that has often been minimized in the ongoing art historians' debate over Brunelleschi's method. This is the considerable discussion of perspective drawing already mentioned by Brunelleschi's near contemporary and fellow artisan Antonio Averlino called Filarete in his treatise on architecture written in the early 1460s, twenty years

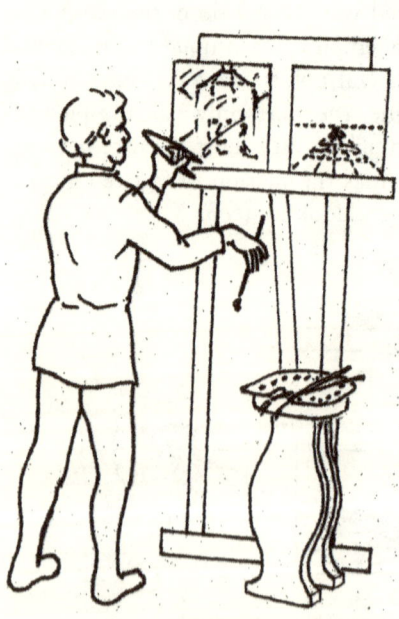

36. Brunelleschi transferring mirror image to his panel. Drawing by author.

before Manetti's biography of Brunelleschi. Filarete, who may well have been an eyewitness to Brunelleschi's original demonstrations, specifically recalls him as the "discoverer" of perspective. Even though Filarete was just as familiar with Alberti's writings, which he also cited, in particular the method for projecting a foreshortened pavement (*pavimento*), he then stated unequivocally:

I believe that Pippo di Ser Brunellesco the Florentine discovered the method of making this plane in this way. It was certainly a subtle and beautiful thing to discover by rule from what the mirror shows you. . . .

If you should desire to portray something in an easier way, take a mirror and hold it in front of the thing you want to do. Look in it and you will see the outlines of the thing more easily, whatever is closer to you, and that which is farther away will appear to diminish (*diminuire*). Truly I think it was by this method that Pippo di Ser Brunellesco discovered this perspective, which was not used in other times, for even though the intellects [of the ancients] were very subtle and sharp, they never used perspective. Even though they exercised good judgment in their works, they did not locate things on the plane in this way and with these rules. . . . Giotto and many others . . . did not use these measures, foreshortening (*scorci*), or all the things they should have. . . . Look at their buildings, for sometimes the figures are almost as large as the houses. Many times they also show the above and below of a thing at the same time.[12]

In the same folio page, Filarete went on to explain how to draw buildings in elevation according to the method he had just ascribed to Brunelleschi. The author says nothing that could be construed as "orthographic projection." Indeed, he detailed a special means for drawing polygonal, including "octagonal," buildings by setting them out on a foreshortened trapezoidal ground plane, which I consider the real "smoking gun" clue that he had indeed seen Brunelleschi's picture of the Baptistery. Here, in slight paraphrase, are Filarete's instructions:

You have understood about square buildings. If you want to make round ones . . . octagonal, hexagonal, or polygonal, you must first make a circle on your [trapezoidal] plane in . . . [elliptical] form . . . and then [divide it into star-like segments]. From every point on the star raise a perpendicular line. You will make as many [sides of the building] as there are lines in the star. You will need to draw these . . . [sides] to two points; that is, you must place another point on the centric line in addition to the first. If you make this building in the middle [as seen frontally], then place [the points, one each] at either side as far out as seems sufficient for your building to [appear natural].[13]

Brunelleschi, of course, worked out his perspective composition on a more spacious surface than in my illustration 35. He would certainly have allowed himself extended room in which to draw the distant transverse rows receding ever closer together all the way to the certification point. On this detailed foreshortened grid, Brunelleschi would have laid out a precise octagonal plan (as in ill. 59, chapter 11), from which he could easily have elevated the vertical sides of the Baptistery just as Filarete described.

CHAPTER EIGHT

Brunelleschi's Second Perspective Panel

Antonio Manetti went on to describe Brunelleschi's second perspective picture:

> He made in perspective the piazza of the Palazzo dei Signori [Palazzo Vecchio] in Florence together with all that is in front of it and around it that is encompassed by the eye when one stands outside the piazza. . . . [from a corner point of view looking south and east] . . . From that position two entire facades—the west and the north—of the Palazzo dei Signori can be seen. It is marvelous to see, with all the objects the eye absorbs in that place, what appears. . . . One might ask at this point why, since it was a perspective, he did not make that aperture for the eye in this painting as he did in the small panel of the Duomo of San Giovanni? The reason that he did not was because the panel for such a large piazza had to be large enough to set down all those many diverse objects, thus it could not be held up with one hand while holding the mirror with the other hand like the San Giovanni panel; no matter how far it is extended a man's arm is not sufficiently long or sufficiently strong to hold the mirror opposite the point with its distance. He left it to the spectator's judgment as is done in paintings by other artists, even though at times this is not discerning. And where in the San Giovanni panel he had placed burnished silver, here he cut away the panel in the area above the buildings represented, and took it to a spot in which he could observe it with the natural atmosphere above the buildings.[1]

Conventionally, it has been understood that Brunelleschi's second picture was an example of what in present-day practice is called "two-point" or

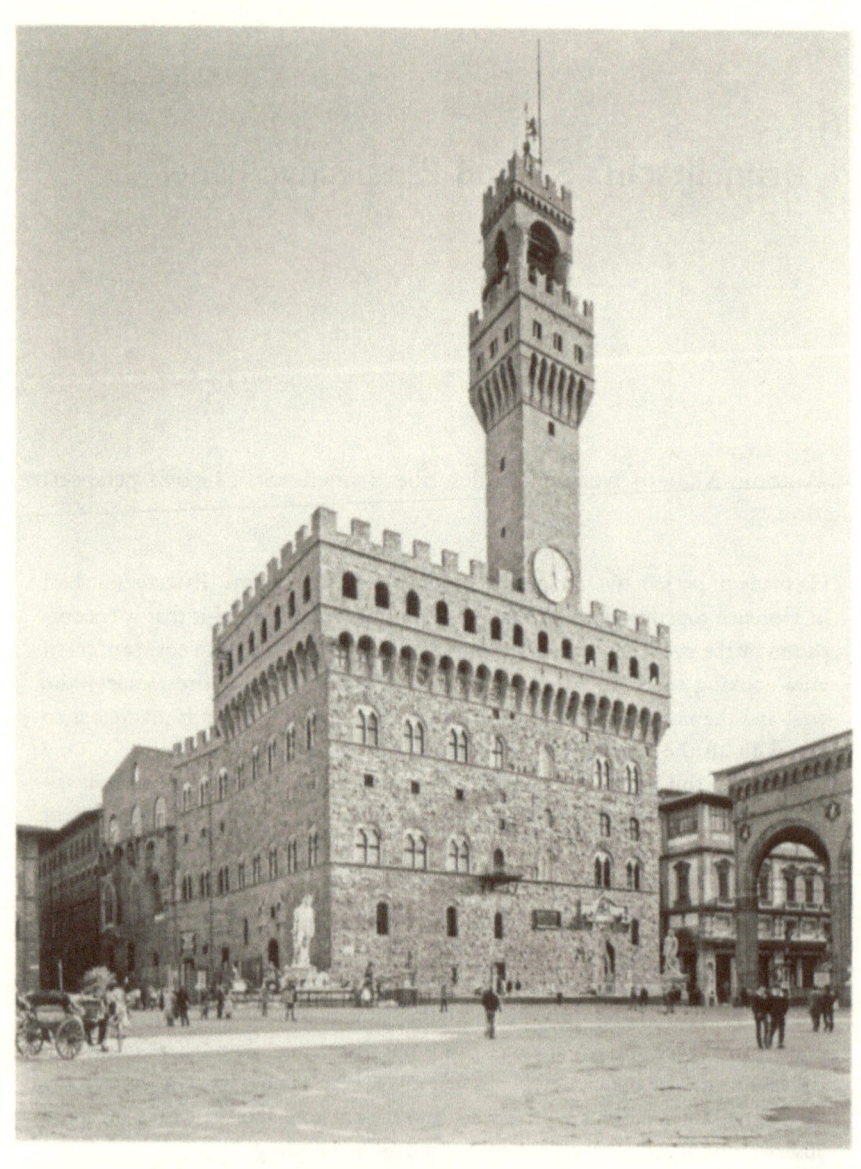

37. Palazzo Vecchio, Florence. Photo: Ralph Lieberman.

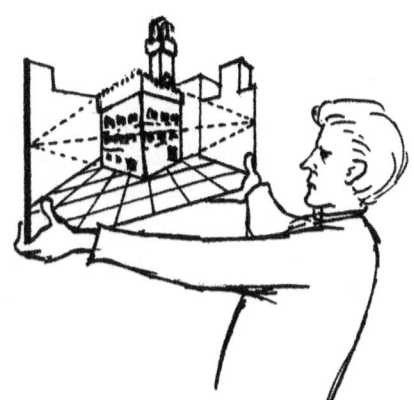

38. Brunelleschi viewing his second perspective panel. Drawing by author.

"oblique" perspective (ill. 38), while the Baptistery panel should be considered "one-point" or "frontal" perspective. Again, I argue that such modern-day terminology does not necessarily apply to Brunelleschi's method. Before attempting to assess Brunelleschi's second picture here, however, a brief description of the basic differences between "frontal" and "oblique" perspective will be helpful.

If, in the phenomenal visual world, an observer stands between the parallel rails of a set of train tracks looking at them straight ahead on level ground, they will seem to converge, literally to come together and "vanish" at the far distant horizon directly in front of where he or she is facing. If the same observer is a trained artist and draws a picture from this fixed position, it will show the tracks converging on a single "vanishing point" in the center, on a line with the horizon, which in effect is the same as the eye level of the observing artist. Such a picture is said to be in "one-point" or "frontal" perspective.

If, however, one views, say, a quadrangular building from its corner on level ground with its two visible sides running off diagonally from the observer's corner position, the top and bottom edges of each angled side will appear to converge on separate "vanishing points," left and right respectively, but still at the same eye level. If an artist draws a picture of such a corner-viewed building, it will be said to be in "two-point" or "oblique" perspective.

The descriptions above are oversimplified, to be sure. In fact, true perspective pictures can show both frontal and oblique views mixed together, with as many objects set at many different heights and angles as the artist wishes, and thus have any number of "vanishing points." Indeed, once the basic geometry of linear perspective is mastered, all sorts of exploitations of the system are possible. Think of those curious anamorphoses that so intrigued Renaissance painters such as Hans Holbein (the haunting skull hidden in his 1533 *Ambassadors*, for example), and the amusing perspective anomalies in the more recent art of M. C. Escher and René Magritte. Nevertheless, every true perspective image does presume a logical viewpoint even when the subject is illogical. No matter how complex, each perspective scene retains a clue that will indicate the artist's original viewing angle and eye level from whence that individual detail was imagined.

Such is the nature of artistic freedom. However, one is still able to distinguish deliberate perspective distortion by otherwise knowledgeable artists, from the innocent mistakes in paintings by naïve artists ignorant of linear perspective. We are reminded of the delightful dialogue described by Plato in his *Lesser Hippias*, where Socrates teases his Sophist friend concerning which is "better," to do wrong voluntarily or to do wrong involuntarily. When Hippias supports the latter, Socrates confounds him with paradoxical examples—which might even have included perspective artists. I have often challenged my students to look at any Renaissance painting and determine, as I believe I can, which was done by an artist who understood perspective but deliberately did not follow the rule, and which was done by a naïve artist who did not know the rule.

We return to Brunelleschi's second perspective picture. First, why might he have chosen the "Palace of the Signori" as the subject? One purely mathematical reason was surely that he had quickly realized, as does anyone who learns perspective drawing even today, that the geometry of the "one-point" frontal system, which he divined from catoptrics and applied to his Baptistery painting, is exactly the same as that which one needs to create the illusion of "two-" or "multiple-point" oblique perspective. Nothing really changes regarding the artist's initial eye-point in the center (even in his second picture as affirmed by Manetti) except that the subject rotates within the picture from a frontal to an oblique corner view. In fact, the octagonal Baptistery building exhibits both frontal and oblique facades simultaneously. Its form is actually that of a cube with each of its corners sliced off at forty-five-degree angles. Even though conventionally viewed from the front, one could indeed think of the Baptistery as contained within a cube on a corner rotated forty-five degrees.

In any case, as Brunelleschi traced out his first picture, he had to be aware that the top and bottom edges of the two diagonal walls flanking the front façade of this structure converged to separate points left and right, but again at the same level as the little centric hole. Having discerned all this, Brunelleschi would have quickly realized that the Palazzo offered an obvious second challenge, especially because, unlike the Baptistery beheld frontally by onlookers leaving the Cathedral, the Palazzo was, on the other hand, most commonly observed by onlookers approaching from a street corner that only permitted an angled view.

It is therefore probable that Brunelleschi applied exactly the same underlying construction drawing for depicting this building as I argue he did for

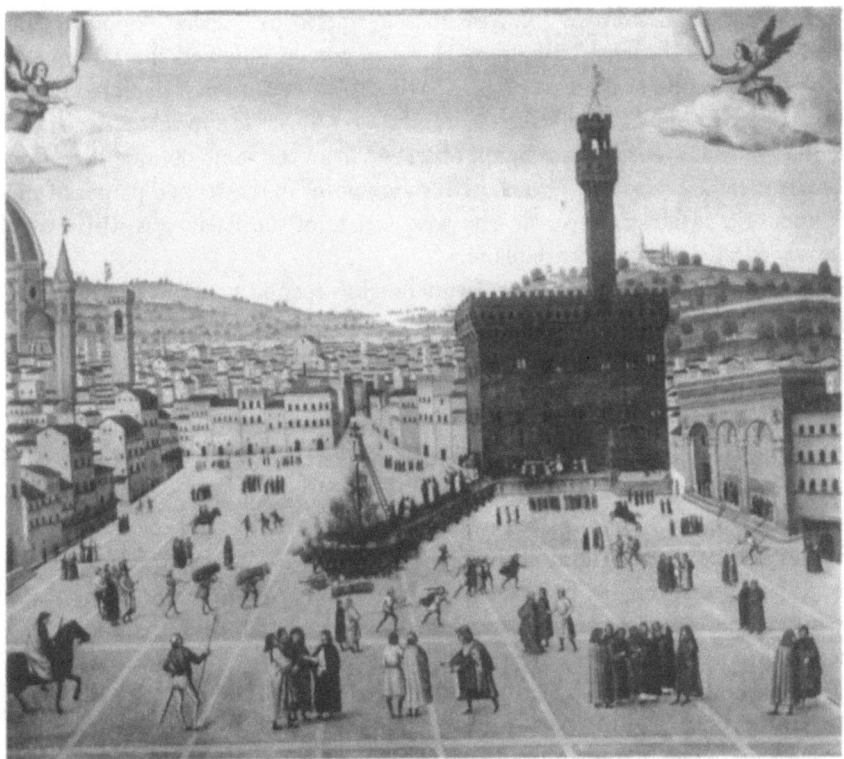

39. Anonymous artist, *The Execution of Savonarola*, San Marco Convent, Florence, ca. 1500. Photo: Alinari/Art Resource, N.Y.

the Baptistery. In fact, there exists a painting, ca. 1500, in the San Marco Museum, Florence, by an unknown Florentine artist ostensibly depicting the *Execution of Savonarola*, which historically occurred on the piazza just to the northwest of the same Palazzo Vecchio (ill. 39).

The gruesome scene is taking place on a gridded piazza projected in ordinary, but somewhat inaccurate, "one-point" perspective, very much like so many iconic Renaissance paintings from the fourteenth through fifteenth centuries (this is a good late example, by the way, where the artist knew something of the rules but still deliberately ignored them for traditional compositional reasons). Nonetheless, the ruled pavement blocks—which may have been only imagined; there's no evidence that the piazza looked like this in 1500—do appear to converge, more or less, on a single point near the upper center of the wide picture. In fact, Manetti, in speaking of Brunelleschi's second panel, mentioned likewise that it was to be viewed "opposite the point with its distance," implying that this image followed the same construction method as the first. While Savonarola's execution is depicted in the midst of the piazza, the Palazzo Vecchio is shown looming to the right, facing west and north, once again observed from the same oblique direction Manetti remembered as Brunelleschi's viewpoint in the second panel. In the Savonarola painting, however, the west façade of the Palazzo is still shown nearly parallel to the picture plane.

Again according to Manetti, Brunelleschi's second panel was larger than the first, and must have included many of the same landmarks viewed from the northwest corner of the piazza, but all at human eye level, rather than from higher up as in the Savonarola scene. In any case, I hypothesize that Brunelleschi may simply have extended the baseline of his original gridded trapezoid to the two right and left "distance points," each a full half *braccio* from the centric certification point, thus doubling the size of the first perspective panel. He might then just as easily have delineated the Palazzo off center to the south (right) so that, even though the western façade of the building would remain parallel to the baseline, the oblique northern façade whose edges must converge toward the horizon would still remain visible—again, just as Manetti remembered (ill. 40).

Like the Baptistery, the Palazzo Vecchio was also a building of great symbolic importance in fifteenth-century Florence. Being the center of municipal government, it represented the city's republican *corporalitas*, just as the Baptistery (and adjacent Cathedral) stood for its moral *spiritualitas*. Brunelleschi actually lived in the Palazzo Vecchio for two months, May and June 1425, when, as already noted, he served as prior, one of the political representatives of his home precinct, the *quartiere* of San Giovanni.

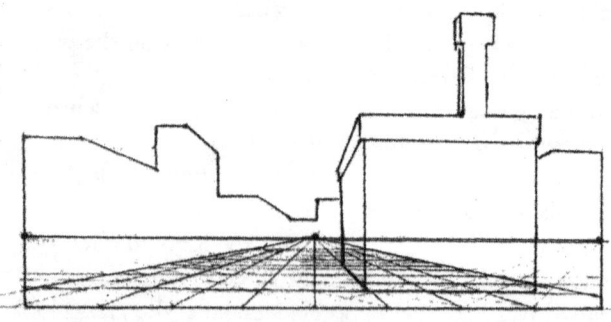

40. Brunelleschi's second perspective panel layout. Hypothetical reconstruction by author.

Antonino, as usual, had much to say in his *Summa* about civic government and the duties of its *rectores civitatis*, as he called all political leaders but not specifically in reference to Florence. He was more concerned about the vicissitudes of mundane civil law generally, and the problems all mortal humans share in rendering justice equally and fairly.[2] Perhaps, Brunelleschi, who was well known in his time as a clever practical joker, intended to emphasize the angled view of the old Palazzo also as a witty comment on the squabbling debates (in which he participated) that no doubt went on inside the building, signifying, like an oblique light ray striking a mirror, the "weaker" temporality of government by living men, compared with the "stronger" perpendicular power of the eternal Church.

Allow me now to review the basic optical principles that Brunelleschi applied to the construction drawings before painting both of his two panels. As already stressed, we can safely infer that Brunelleschi's most important rule was that any picture in proper perspective must, like a mirror image, show a calculable axial relationship between the artist/observer's eye and the "certified" center of the depicted virtual space. His second critical rule was that the phenomenal convergence illusion always be at a level with the viewer's sensible horizon, which every human being experiences in actual vision but which even such precocious painters as Giotto and the Sienese Lorenzetti brothers did not realize before Brunelleschi could be represented in a picture with optical precision. These two rules made it possible for artists to depict the correct illusionary diminution of objects further distant in the virtual space of the picture, reproducing the same relative size difference between objects in the foreground and those in the background, *di naturale* ("according to nature"), as sixteenth-century Italian critics liked to boast.

The third rule, concerning the representation of human figures, is not immediately apparent in Brunelleschi's two paintings as far as we know (there is no mention by Manetti of figures in either painting). It was nonetheless implicit in both. If people were to be depicted in true Brunelleschian perspective, so long as they are intended to be as tall (more or less) as the artist/observer, and standing on the same virtual ground level within the picture as the artist/observer stands in real space without, they should have their painted heads at the same level on the same horizon line as the original "certification" point; that is, the artist's own eye level. In any case, Brunelleschi seems to have passed on this optical principle to fellow artists in Florence who did apply it as we shall see, in several early fifteenth-century paintings, again years before Alberti inscribed the rule in 1435/6.

In conclusion, Brunelleschi's perspective was significantly different from whatever techniques were used by painters before. Indeed, no matter which of the methods that modern scholars have hypothesized as Brunelleschi's original, including Krautheimer's often-cited orthographic projection, something dramatic happened to the art of painting around 1425 in Florence, otherwise the works that followed would have looked no different than the hundreds of nongeometric "empirical perspectives" painted in Europe during the thirteenth and fourteenth centuries. Until Brunelleschi there was never a consistent, geometric-optical basis for these coincidences, although some artists, such as the intellectual Giotto, may have had considerable knowledge of *perspectiva naturalis* as a purely visual science. As I have argued, it was certainly "in the air" in the fourteenth century, but, just as nobody before Brunelleschi could figure out how to place the dome over the Cathedral (embarrassingly open-roofed since it had been begun more than a century before), nobody in the same prior period had made such an effort to connect directly the optical principles of mirror reflection to the creation of an illusionistic picture.

Prometheus Unbound! What Brunelleschi had just wrested from the Creator's closet, and was now about to share with fellow artists, might indeed be likened to that other famous Olympian thievery, granting to mortal man the heretofore sacred privilege of imaging nature just as God himself projected it from his own divine eye.

CHAPTER NINE

Brunelleschi's Heritage: Masaccio's *Trinity*

Giorgio Vasari, in his already cited biography of Brunelleschi, went on to describe how the inventor of *perspectiva artificialis* taught his new system "in particular" to his young friend, Masaccio (1401–28?), whose early master-piece, the nearly twenty-foot tall *Trinity* fresco (ill. 41), restored after World War II to its original location on the left nave wall of the Dominican Church of Santa Maria Novella, Florence, is in all likelihood the very first monumental picture to exhibit Brunelleschi's rules in a prominent public place.

The painting presents an extraordinary illusion of an architecturally or-nate vaulted chapel receding into the wall "as though it were a hole," as Vasari remarked.[1] Quite forward in this virtual space just within a fictive monumental entrance archway (very reminiscent of Brunelleschi's actual ar-chitecture) the artist painted a prominent image of Christ on the cross with God the Father standing higher behind. Between them is the smaller image of the Holy Ghost in the likeness of a dove. These three holy figures stacked together in this order have constituted the canonical iconographic represen-tation of the Holy Trinity since time immemorial, although traditionally God the Father was often depicted much larger in size than both Jesus and the cross (ill. 42).

What is so amazing, however, is that this subject, the most esoteric of mysteries in all Christianity, should be the first example, perhaps the first public exhibition, of Brunelleschi's new perspective. One can only assume that Masaccio and his patrons hoped that by painting this sacred subject ac-cording to the startling new rules of linear perspective, viewers would behold

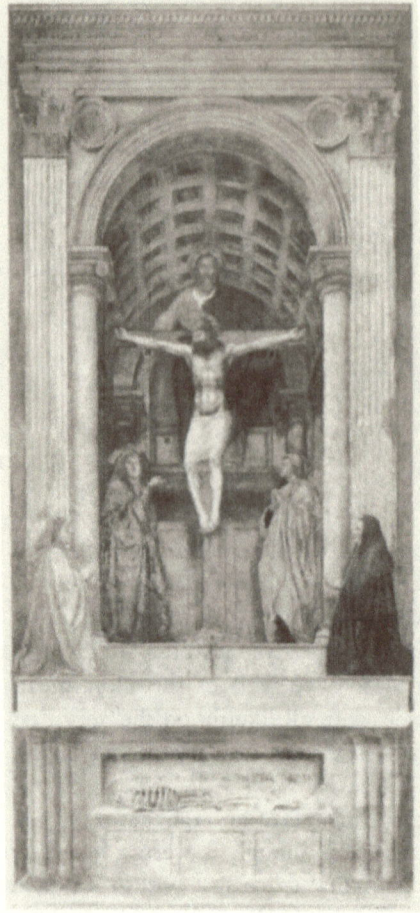

41. Masaccio, *Trinity*, Santa Maria Novella Church, Florence, 1425. Photo: Gabinetto Fotografico del Polo Museale Fiorentino.

it "just like [the prophets] see God or his divine mysteries behind the images and likenesses of sensible things"; that is, as a mirror reflection literally of the "real" Trinity in heaven, just as Antonino preached in his sermon on Corinthians 1:13, which we reviewed earlier.

This very metaphysical subject was thus intended to be represented in human scale, according to the laws of human vision, showing the immortals as if they had come down from heaven in the *species* of living humans to meet their mortal worshippers on earth. Masaccio wanted them also to be understood not only as the same size but even illusionistically reduced in size relative to their near or far positions in the perspective depth of the scene, just as he had learned from Brunelleschi.

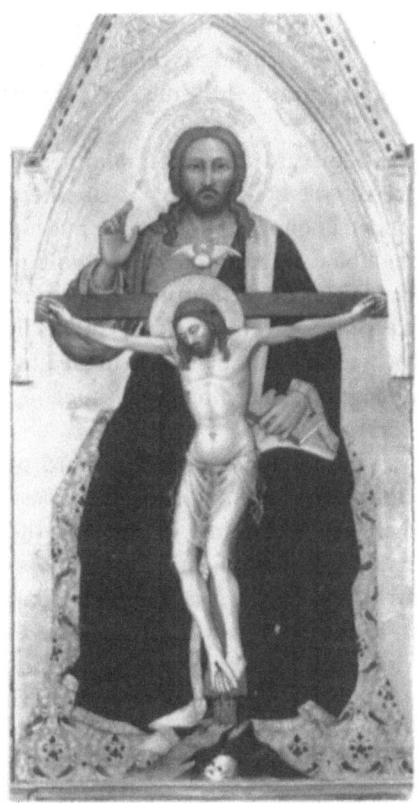

42. Nardo di Cione, central panel of *The Trinity*, Ghiberti Altarpiece, Accademia, Florence, 1365. Fototeca Berenson, Villa I Tatti, Florence. Photo: Antonio Quattrone.

In Masaccio's modified version of the *Trinity*, God the Father was not only reduced to the same physical height as all the other figures in this scene, but, because he stands furthest within the picture space, he must appear the shortest of all as would be the illusion of him in proper optical perspective. Masaccio hoped to lessen the effect of this problem by having God no longer seated on the Throne of Grace as was traditional but represented standing fully upright, in order, according to the iconic tradition, that he visibly tower over Christ, and with his outstretched hands support the arms of the cross. God is thus shown somewhat awkwardly standing, almost teetering, on a narrow ledge protruding from the Ark of the Covenant, Masaccio's necessitated iconographical alternative, filling the back space of the fictive chapel.

Below and on either side of the cross within that virtual space to our left and right stand an aged-looking Virgin Mary and a youthful Saint John the Evangelist, this time appearing even shorter in size than both Jesus and God the Father because of their lowered positions within the painted chapel. The viewer standing in front and slightly below should not see their feet occluded by the raised threshold of the fictive opening. Nevertheless, if their feet were visible, the two saints from tip to toe would be nearly the same size as Jesus and God the Father.

On that painted altar, seeming to extend forward from the monumental chapel entrance as if projected into the actual aisle of the church itself where the viewer stands, the artist depicted the mortal donors, appropriately outside the sacred space of the chapel itself. As shall be demonstrated, these two figures, the nearest to the viewer, while appearing to be the least in size because of their kneeling positions, are actually the only ones among all the others in the fresco whose full measure equals Masaccio's modular human height, more or less three *braccia*—that is, if the length of their horizontal kneeling legs were added to that of their upright torsos.

The *braccio*, as already mentioned, was the then Florentine standard of measure, being just shy of twenty-three inches in length.[2] Indeed, three *braccia* or roughly five feet nine inches was believed at the time to represent the "average height of a man's body," as Alberti would confirm in 1435. It was also the proverbial height of Jesus, and the reason for the *braccio* measurement in the first place. Back in the early 1200s, so the story goes, a Tuscan delegation of merchants headed by the great Pisan mathematician Leonardo Fibonacci, seeking to establish an absolute and inviolate standard of measurement to be used in the burgeoning cloth trade, visited Jerusalem and the Church of the Holy Sepulcher. There, they measured the length of the tomb of Christ and divided the result by the sacred number three. Hence was established the standard *braccio*, literally, a "forearm's length." Jane Aiken has shown that Masaccio pretty much depended on this *braccio* standard as he laid out his fresco on the wall.[3]

In any event, the *Trinity* was probably painted around 1425, a little later perhaps, but certainly not before. Some scholars claim it was completed nearer to 1428, on the grounds that the perspective is too sophisticated to have been the earliest example of Masaccio's debt to Brunelleschi. I would maintain the opposite, that the painting shows not only Brunelleschi's new and optically rational ideas with exuberant enthusiasm but some of the unnecessary and even incorrect ones that were tried out here for the first time, didn't work well, and so were never repeated again in the artist's later work.[4]

For example, Masaccio intended to follow to the letter Brunelleschi's rule, described by Manetti, that "the painter needs to presuppose a single place whence his picture is to be seen," positioning his central certification point not only at a proper distance from the picture surface but even to his own actual height above the real ground where he stood as painter, just about five feet nine inches above the floor of the church. Unfortunately, whatever evidence there was in the painting where the original certification point should have been marked is in the very area most damaged. The lower portion of the fresco where this point must have been was completely covered behind a masonry altar when Giorgio Vasari restored the church in the sixteenth century. During the 1860s the still discernable upper portion was detached and ripped off the wall at this very juncture and then remounted in another location. After World War II, during further restoration of the church, the altar blocking the original bottom half was finally removed, exposing for the first time in nearly four hundred years that lower portion of the fresco still in situ. The upper portion, since moved elsewhere, was then detached once more and refitted to its approximate original place above the lower, with a modern reconnecting patch in-painted at the badly damaged juncture.[5]

Nevertheless, the reattached fresco neatly revealed a clever clue as to exactly where Masaccio did intend to place his certification point. Below the fictive chapel and extending forward from under the altar where the depicted donors kneel, Masaccio painted a tomb on which a human skeleton reclines, stretching from end to end in the middle of its perspectivally receding slab-cover surface. However, the skeleton's true size must be determined not by measuring its length from where it appears in the virtual depth of the picture, but from where it would be if positioned along the tomb cover's front edge projected forward against the picture plane. Here, the true intended length of the skeleton measures just a few centimeters more than three Florentine *braccia*. In other words, if this supine skeletal figure were pulled forward, thus gaining "height" in reverse perspective, and then rotated ninety degrees into a position standing on the real church floor, its "eye" level would indicate the perspective horizon of the whole picture.[6] Already by the fifteenth century in Florence, three *braccia* had come to represent the height of "Everyman," if not the actual head height or eye level of Masaccio himself. The artist further emphasized the mortal, moral message of this skeletal Everyman by inscribing the following elegiacal couplet (loosely translated to keep the rhyme scheme) just above the bier: "As you are now so once was me; as I am now so will you be" (*Io fu gia quel che voi siete e quell chio son voi anco sarete*).

A five-feet-nine-inch height more or less above the church floor just reaches the upper edge of the painted altar where the donors kneel. Even though most of this part of the fresco has been restored in modern times, one may still be sure that this is the place where the certification point would have been, for the simple reason that the original edge of this fictive altar surface appears only as a straight line. Just enough of the original fresco in this area remained attached to the upper portion to be certain this is so, even after the lower portion was separated. This is clearly apparent in illustration 43, a photogram showing where original incision marks in the upper portion of Masaccio's *Trinity* were discovered during recent restorations.

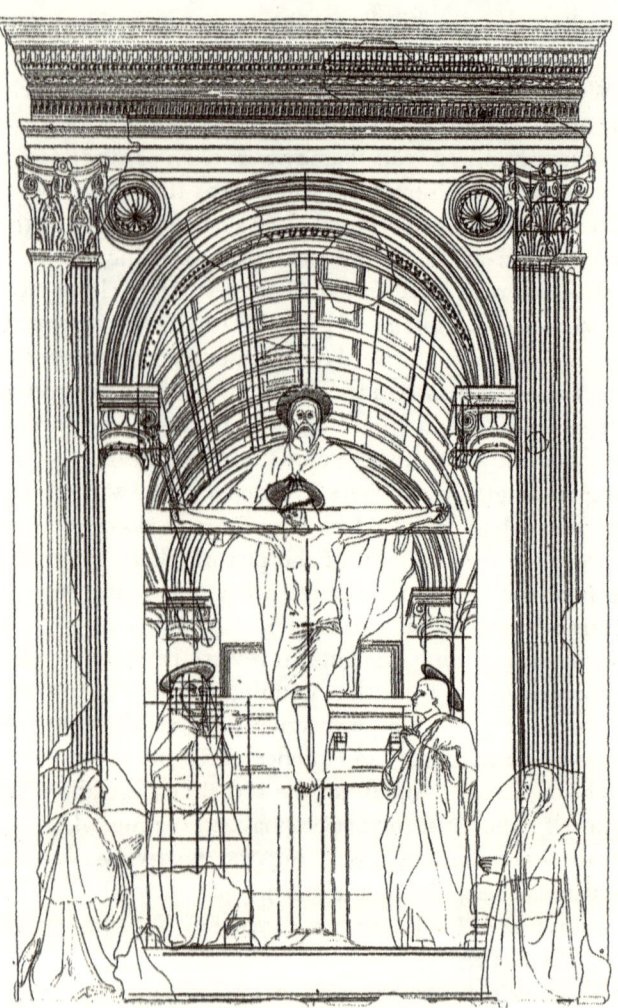

43. Reconstruction of Masaccio's *Trinity*, showing traces of original layout lines incised in the *intonaco*. Photogram from *La Trinità di Masaccio: Il restauro dell'anno Duemila*, edited by Cristina Danti (Florence: Edifir, 2002), 13.

In actual vision, as Brunelleschi must have also discovered, any flat horizontal surface looked at on edge and at the same level as the observer's eye-point will appear as a straight line, with the surface itself hidden from sight (ill. 44).

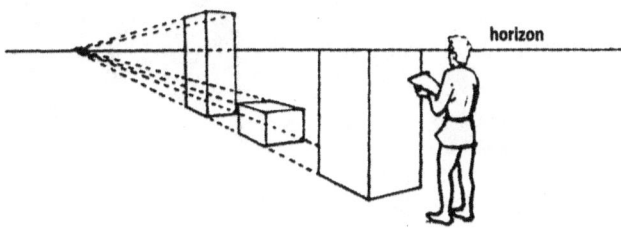

44. Brunelleschi observing the coincidence between his eye level and the horizon. Drawing by author.

Thus, Masaccio clearly meant this thin edge underneath his depicted donors to represent the average human eye level, the horizon, in the middle of which between the two figures he should place his point of certification.[7] To this he would then have directed the orthogonals, the receding edges of his fictive vault, which, if the structure were real, must appear to the eye to be converging on that point (ill. 45).

Close examination of the middle-aged figure of the Virgin Mary in the *Trinity* reveals another interesting detail that demonstrates how Masaccio applied Brunelleschi's system to his large wall fresco. Under raking light, one is able to discern an intaglio grid of squares lightly incised in the fresco surface where her face is painted (ill. 46). This surely indicates that the artist had first made a smaller drawing of the figure on a sheet of squared graph paper and then transferred it point by point to the larger squared grid scratched in the fresco.

It's also important to remember that when the upper part of the fresco was detached from the wall in the nineteenth century, no *sinopia* was recorded, that is, the characteristic underdrawing almost always found under the thin *intonaco*, the ultimate layer of plaster on which the artist paints his final picture. It would thus appear that Masaccio laid out his perspective scheme on top of the *intonaco* after it was already applied. Moreover, numerous lines are disclosed incised on this surface by raking light, including long diagonals by which the artist directed the orthogonals of the architecture to converge on the certification point, and a series of looping curves indicating where he

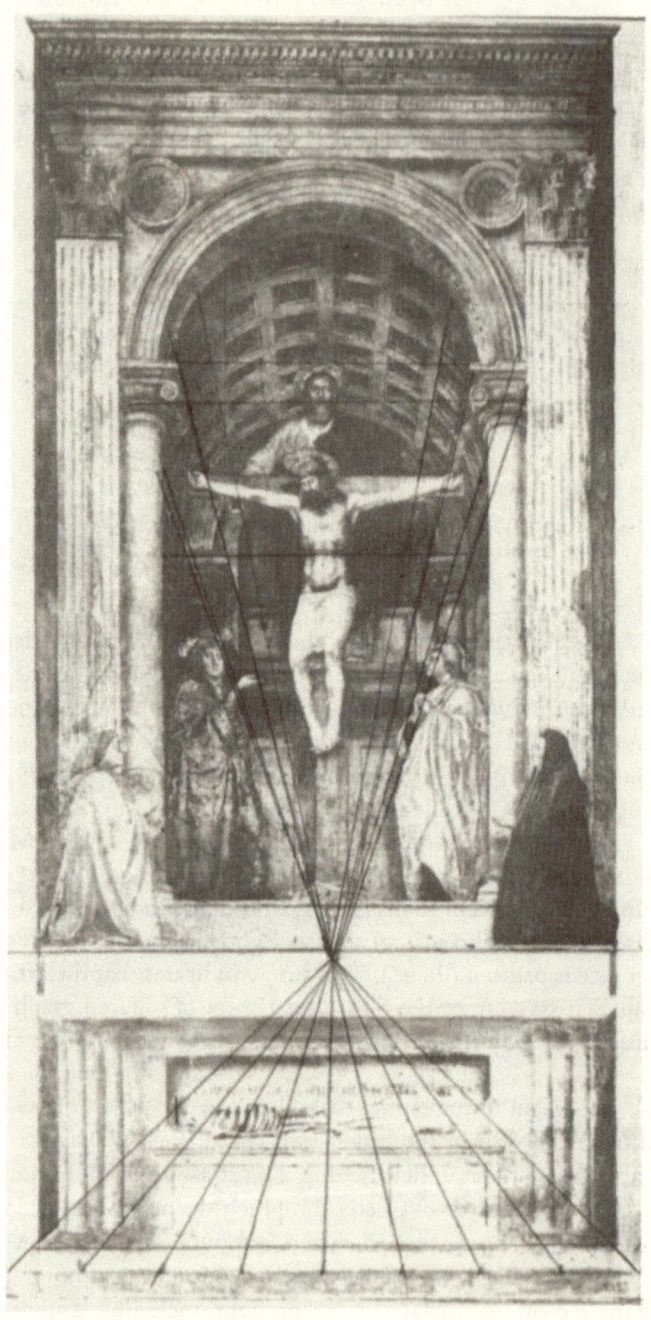

45. Reconstruction by author of Masaccio's *Trinity* showing point of perspective convergence.

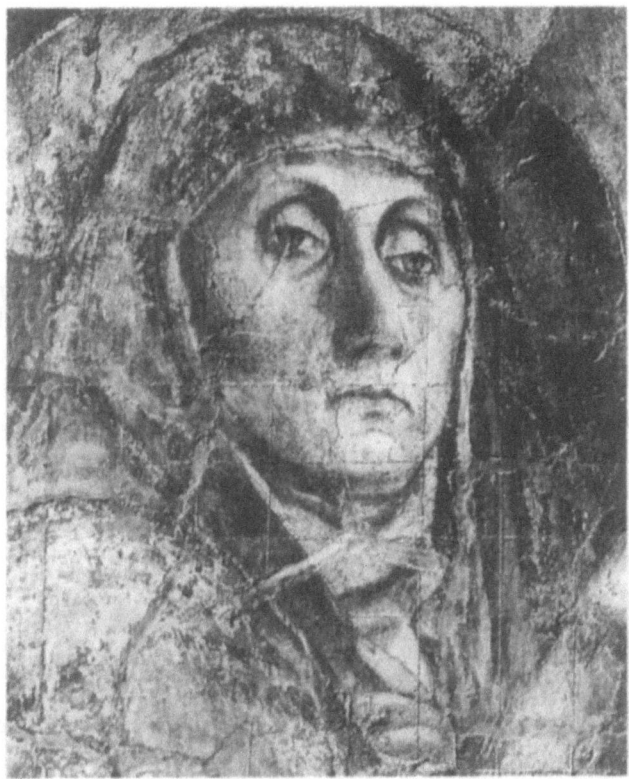

46. Detail of Masaccio's *Trinity* showing incised grid on the
Madonna's face. Fototeca Berenson, Villa I Tatti, Florence.
Photo: Luigi Artini.

should paint the arches of his fictive vault. Illustration 43 is a reconstruction
of all the incised lines still traceable on the surface of the detached and repositioned original upper portion.

One must conclude that Masaccio laid up his entire fresco in this manner;
that is, by working out the perspective construction initially in small scale on
grid-lined paper and then blowing it up, detail by detail, onto the larger
squares incised in the plaster *intonaco*.[8] The artist also may have inscribed
numbers and letters on "strips of parchment" tacked up on the sides and
along the top of the fresco, just like the cartographic-style notations that Manetti said "Filippo alone understood" while measuring antique buildings in
Rome. Such a coordinate system would have helped also to transfer his
smaller drawing to the full-size painting. All he needed to do was to make

sure that the various elements of his perspective transfer properly aligned along converging orthogonals, each leading to the single central certification point on his prefixed horizon.

On the other hand, there will always be fudging in any such transfer system, due to miscalculation, a change of mind, or just plain readjustment for aesthetic reasons (not to mention the damage suffered by the fresco during its several detachments and unsure restorations), all of which make it impossible to recapitulate by modern measurement just exactly how the whole perspective scheme was planned. Indeed, when Masaccio blew up the central Crucifixion detail from the scale drawing, he seems to have painted in the figures of Jesus and God the Father freehand, as if from a slightly higher viewpoint than the fictive vault itself so that they should appear more frontal in the usual iconic tradition.

Such a mixture of freehand and ruled perspective was certainly shown to be the case with the *Trinity* when, in 1996, science historian Judith V. Field had the rare opportunity to make measurements directly upon the fresco surface in situ. Nevertheless, even though she found no specific evidence, Masaccio must surely have at least begun his construction by drawing a traditional gridded trapezoid ground plan such as was also a common artistic convention for the presentation of iconic subject matter at the time. The arched and coffered vault in the painting definitely had to be projected from just such a gridded plane, although Field again found no incised lines to indicate how that may have been done.[9]

Among Field's more precise findings, however, is the height of God the Father as Masaccio rendered him standing behind the cross. He is just 155 centimeters, or about five-feet-one-inch tall. Unfortunately, she did not measure the parallel height of Christ because she assumed that the artist intended him to be the canonical three *braccia*, or 175.08 centimeters tall. This cannot be so, even though the fully visible figure of Jesus in front and center of the scene seems to loom even larger than God. No matter; for all intents and purposes he is the same size as the latter. Although the skeleton on the tomb below was the intended height module for all the figures in the fresco, neither God nor Jesus, since they stood further within the picture space, could be perceived in normal perspective vision as fully three *braccia* tall.

Once again, we refer to Fra Antonino. It's quite possible, and perhaps even likely, that Antonino had seen and thought about Masaccio's *Trinity* when, later as Archbishop of Florence, he wrote in his *Summa* his only passage specifically about the responsibilities of painters in Florence and the didactic role of their paintings (which he termed *historiae*, "histories"). Art historian Creighton Gilbert commented on this unique passage several decades ago, pointing out that the Archbishop, without mentioning names, was actually favoring Masaccio's austere representations of sacred subjects, while disparaging the currently fashionable International Gothic style (a classic contemporaneous Florentine example of which was Gentile da Fabriano's 1423 *Adoration of the Magi*; see ill. 75). Such frivolous works, Antonino opined, did not "excite devotion but only laughter and vanity" (*quae non valent ad devotionem excitandem, sed risum et vanitatem*),[10] worrying further that some artists of this persuasion even depicted "oddities" (*curiosa*) such as a three-headed representation of the Trinity, which is "monstrous in the nature of things" (*monstrum est in rerum natura*).[11]

We are presented here with both an iconic problem and an ingenious response. Although Masaccio maintained an "earthly" perspective on this very sacred scene, he also intended to manipulate his new Brunelleschian rules, so that they still communicated some of the old hierarchical symbolism.

I suggest that Masaccio was already aware of Antonino's prejudices when he began constructing his own Trinitarian imagery, especially in a Dominican church.[12] In the traditional depiction of this subject, God the Father was usually shown as an unnatural giant compared to the other figures. This was, of course, before Brunelleschi added his precise optical rules to correct his "mirror of nature." Might not then Masaccio, with both Brunelleschi's and Antonino's blessing, have realized that in this new context it would be "monstrous" to depict God so huge, especially when his *species* was being transformed by divine optical and catoptrical reflection into earthly form?

On the other hand, Florence in the 1420s was still a very conservative religious community, and Antonino was already being regarded as its spiritual spokesman. Wouldn't he have wanted also to preserve certain painting traditions that did "excite devotion"? Perhaps urging Masaccio via Brunelleschi to have it both ways in this first optically correct version of the *Trinity*? Whether or not Masaccio conceived the solution himself, he did indeed cleverly arrange the locations of the six figures in such different perspective positions that their relative heights, as painted, remarkably signified their traditional hierarchical sizes and thus their rank in the Great Chain of Being. For instance, while Jesus on the cross (the only completely visible, full-length figure in the painting) certainly catches the viewer's attention most dramatically, God the Father still looms above Jesus's head, implying his superior status.

Then, in order that the similar, nearly equal-sized Virgin Mary and Saint John not compete with the strong visual emphasis on God and Jesus, Masaccio literally lowered them in perspective. Since they stand on the floor of the raised chapel, their feet would be out of sight to any five-foot-nine-inch viewer in front. Hence, they too look considerably smaller than God and Jesus. Finally, the two foreground donors, the only figures nearly the same height as the viewer, were cleverly made to kneel, reducing their sizes so that they appear smaller than all the others, as befitted their humble mortal station at the bottom of this cosmic hierarchy.

Fra Antonino would have been pleased with this extraordinary tour de force. He might even have complimented the artist when he later commented in his *Summa*, quoting Luke 10:23: "Blessed are the eyes which see what you see. . . . Many prophets and kings have desired to see what you see, but they did not see."[13]

Masaccio thoroughly believed, as did Brunelleschi during these early demonstrations of perspectiva artificialis, that the pictorial illusion could only be experienced properly by the viewer standing at the same eye level as the artist stood when the picture was painted. This turned out not to be necessary, as Masaccio himself quickly realized by the time he planned his next commission, the fresco cycle in the Brancacci Chapel. The *Trinity* remains the only monumental public painting of an iconic religious subject by any artist during the fifteenth century where life-size figures are depicted in virtual space just as if they would be seen diminishing perspectivally by a same life-size human observer standing in front whose eye level was coincident with the certification point at the picture surface.

CHAPTER TEN

"Oh, che dolce cosa è questa prospettiva!"

Donatello and Masaccio

The ever quotable Giorgio Vasari tells another quaint story about the Florentine painter Paolo Uccello (1396/7–1475), contemporary of Brunelleschi, Masaccio, and Donatello. Paolo's wife, worrying that he was staying up too late at night working, entreated her husband to come to bed, only to have him ignore her in his total preoccupation with the new art-science. "Oh, what a sweet thing is this perspective!" she heard him muttering to himself in wondrous excitement.[1]

Paolo's famous geometric drawings, like the *Chalice* (ill. 47), were so meticulously measured according to exacting geometric-optical rules that even his friend Donatello cautioned that such obsessive attention was causing him to "leave the substance for the shadow."[2]

Nevertheless, if there is one vivid clue as to when and by whom Brunelleschi's new principles were first used, it is in this very enthusiasm, evident in perspective works executed by other artists even earlier than Paolo's drawing (which is dated rather late: around 1435). Already, a decade earlier, we see this same perspective zeal in certain of their works, datable to 1425 or shortly after, compared to the lack of zeal in works before 1425. I shall call this my "excitement gauge."[3] Unfortunately, it is difficult for modern eyes, so overly accustomed to ever more tricky perspective illusions since the Renaissance, to appreciate how surprised people once were upon first seeing an optical perspective image. Indeed, we have evidence of such sudden amazement from seventeenth-century Asia when the Jesuit missionaries first introduced Western-style perspective pictures to China, Japan, and India.[4] Unlike loud music, which never fails to startle the ears no matter how often replayed, perspective illusions quickly become déjà vu and even boring with repetition, and always needed to be reinvigorated with added novel affects in order to reexcite the eyes.

I would now apply my "excitement gauge" as evidence that not only was early 1425 the date when Brunelleschi first introduced his geometric system, but also to demonstrate that it had nothing to do with "orthographic projection," rather only with an emphasized centric certification point fixed upon a uniform horizon line, and with the figures set upon, or implied as set upon, a geometrically projected gridded floor plane. I also take issue with those scholars, such as John White and James Elkins, who have argued that the advent of linear perspective in Western art was a gradual development, actually beginning in antiquity and slowly gaining traction through the Renaissance, and that Brunelleschi's remarkable optical demonstrations were only a blip in this slow evolutionary process and not a sudden "big bang," as I have proposed.

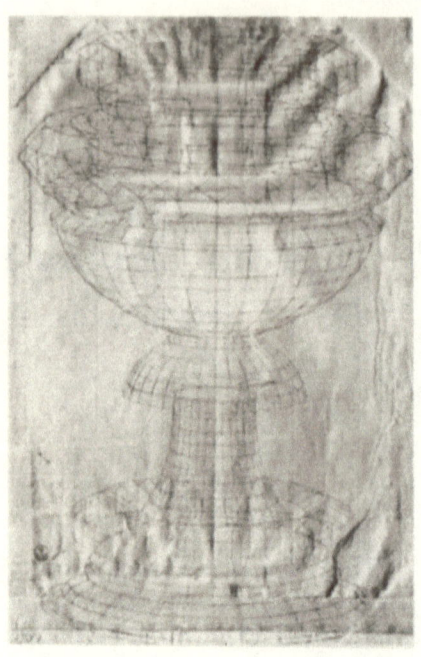

47. Paolo Uccello, perspective drawing of a *Chalice*. Galleria degli Uffizi, Florence, ca. 1435. Photo: Gabinetto Fotografico del Polo Museale Fiorentino.

I submit now two pairs of artworks, each by the same artist and each consisting of a "before" and "after" component. One pair is by Brunelleschi's older companion, Donatello, and the second by his young friend, Masaccio. Applying my impromptu "excitement gauge," I will demonstrate that the two artists did not know Brunelleschi's rules in their "before" examples; that is, prior to 1425, but had learned and applied Brunelleschi's optical geometry exactly and with manifest exuberance by the time they finished their "after" examples in that or the next few years.

My first example of artistic excitement with Brunelleschi's perspective might even have preceded Masaccio's *Trinity* in deserving the honor of being the very first to demonstrate his rules. This is the relief panel depicting the *Feast of Herod* carved for the baptistery font in Siena by Brunelleschi's close friend Donatello, which seems to have been originally begun sometime around 1424 and finally cast in bronze and displayed in 1427. But let me demonstrate by comparing it to a prior work by the same Donatello that has nevertheless often been cited in the modern art historical literature, errone-ously I believe, as the very first example of Brunelleschian linear perspective: the stone relief panel depicting *Saint George and the Dragon* from the taber-nacle dedicated to that saint on the north side of the Florentine Church of Orsanmichele, carved around 1417 (ill. 48).

There is no doubt that Donatello, like Brunelleschi, was much intrigued even at this early date by the science of *perspectiva naturalis*, and this is clearly evident in the remarkable background illusion of a distant forest bathed in a misty light defined here by a very low, compressed relief behind the project-ing figure of Saint George. Art historians have correctly asserted that this is a prescient rendering of optically observed atmospheric perspective.[5] To the right behind the princess whom Saint George is rescuing, there is a bit of an arcade with a few incised diagonal lines demarcating a tile floor that seem to be directed toward a single point more or less level with the maiden's head, nonetheless hardly a confident exhibition of optical linear perspective.

But now compare this hesitant illusion with the same, far more enthusias-tic attempt in the artist's *Feast of Herod*, cast in bronze less than a decade later, which literally bursts with any number of perspective tours de force (ill. 49). Donatello not only dramatically emphasized an isocephalic horizon line with centric certification point but he added four protruding beams above and even a receding niche in the wall below that have no other purpose than to show off the artist's excitement at his ability to create such unprecedented optical effects.

Surely, in this most energetic work, Donatello was parading a new trick, one he had learned either just before or just after Brunelleschi was teaching the same to an equally excited Masaccio. As mentioned, Donatello began the relief around 1424, one of six panels by several artists to be installed around the hexagonal font in the Baptistery below the Sienese Cathedral. At first, he was probably still dependent on the traditional empirical bifocal method for creating the illusion of depth, especially as he etched in the lower portion of his wax preparatory block in order to indicate the squared pavement beneath the narrative scene. John Shearman has pointed out that this detail resembles the similar tiled floor depicted in Ambrogio Lorenzetti's fourteenth-century *Presentation* (ill. 29), which at that time stood just above the Baptistery inside the Cathedral proper.[6]

Within the next year, however, and apparently after having just mastered the new optical perspective rules, Donatello redesigned the upper half of his relief by emphasizing a new horizon line sharply cut and running along the

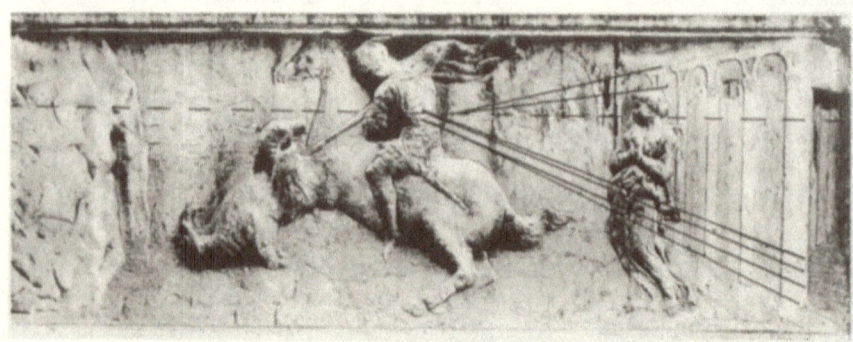

48. Perspective reconstruction by author of Donatello's *St. George and the Dragon*, stone relief from the Church of Orsanmichele, Florence, 1417.

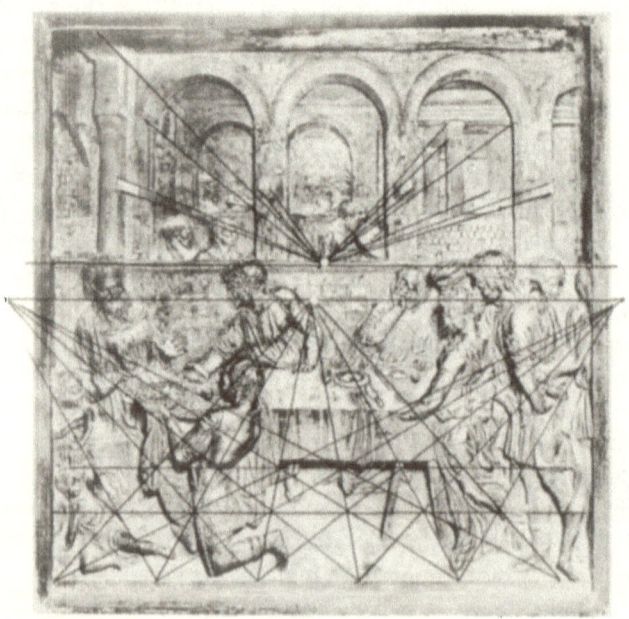

49. Perspective reconstruction by author of Donatello's *The Feast of Herod*, bronze relief from the baptistery font, Cathedral of Siena, 1424–47.

protruding edge of a high wall separating foreground from background. To this level he then raised the heads of the foreground figures, thus compelling the viewer to concentrate on their horrified expressions as they witnessed this grim biblical tragedy.

Certainly, Donatello would have made several drawings on paper before tracing and cutting the details onto a block of wax. Understandably in such a process some of the linear accuracy of the original drawing would be fudged. Like Masaccio, Donatello realized that in order to make his composition ever more dramatic, it might be necessary to modify some of Brunelleschi's rules. In this case, he retained the empirical bifocal gridded floor, but kept it low and separate from his true perspective lest it appear incongruously steep if elevated to the actual certification point of his extraordinary background illusion. Here is another early instance where a great artist, however intrigued by a new set of pictorial principles, did not hesitate to readjust them to suit his own natural style.

We now examine a painted triptych (a three-panel altarpiece) only discovered in the last century that after some dispute is generally attributed to Masaccio. It is dated "April, 1422" from the rural town of San Giovenale di Cascia di Regello near Florence, and it is currently in the Uffizi Gallery (ill. 50).

I pose this beside the same artist's third great masterpiece, the central panel of the similarly formatted so-called Pisa Altar, documented as having been painted in 1426, but which now, separated from its adjacent side-, top-, and bottom-predella parts (some are extant but in other European and American museums), now hangs alone in the National Gallery of Art, London (ill. 51).

Both Masaccio's panels depict the Virgin Mary with Baby Jesus enthroned and two small angels at her feet. Two more little angels kneel behind the throne in the Pisa Altar, as in the San Giovenale triptych. A pair of side extensions to the Pisa panel, now lost, each similarly depicted a pair of larger standing saints left and right as Vasari, who witnessed the original altarpiece before it was broken up, attested.[7]

In the San Giovenale central panel, however, the Virgin's throne rests on a sharply upward sloping trapezoidal floor, vaguely patterned with converging diagonals as was the commonplace empirical convention in 1422, three years before Brunelleschi's demonstrations. Notice, nevertheless, the grid lines are barely visible here, mostly covered up by the robes of the flanking saints. In fact, if one endeavors to track these, as well as the converging edges of the architectural throne, in search of an identifiable certification point, the various diagonals do not meet in conjunction with an equally pronounced

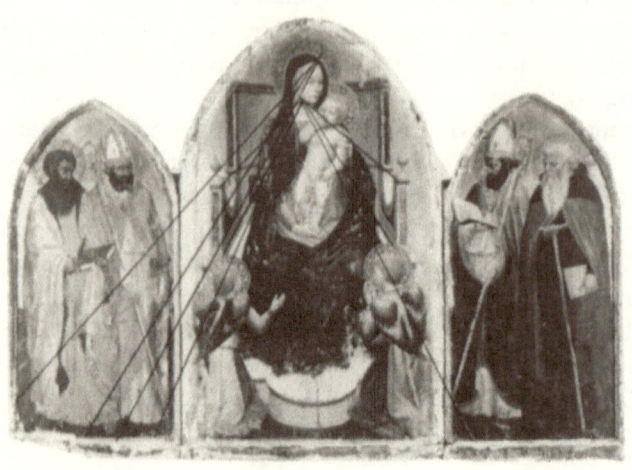

50. Perspective reconstruction by author of Masaccio's *San Giovenale di Cascia di Regello* Madonna, Galleria degli Uffizi, 1422.

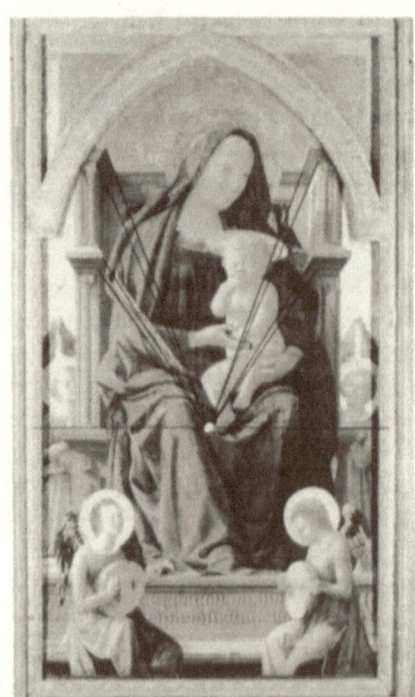

51. Masaccio, *Virgin and Child* (*Pisa Madonna*), 1426. Bought with a contribution from the National Art Collections Fund, 1916. © The National Gallery, London. Perspective reconstruction by author.

horizon line, but rather gather more or less in a cluster around the Virgin's face at the very top of the panel, high above the heads of all the standing and kneeling figures below. Although Masaccio seems to have sensed that the traditional empirical convergence phenomenon might be manipulated to heighten the iconic focus of his sacred subject, he surely knew nothing yet of Brunelleschi's certification point and horizon line principle. In the next chapter, we will see how Masaccio again applied this same iconic emphasis in his Brancacci Chapel *Tribute Money* where he directed all the architectural orthogonals in the picture toward a single centric point on the head of Christ, but this time in firm Brunelleschian conjunction with the uniform horizon.

In no other works of Masaccio prior to the Brancacci Chapel frescos except the *Trinity* is Brunelleschi's 1425 influence more obvious than in the Pisa Altarpiece. First, we must be aware that the figural images here are quite small, not nearly in life-size scale as in the earlier masterpiece. Also the original all-together Pisa altarpiece once stood raised on a regular church altar. Because of that, Masaccio did another clever thing with his new perspective: he established his horizon right at the Virgin's knees with the certification point centered between them, which surely was intended to correspond more or less with the eye height of any celebrant standing before the altar. The effect, of course, was to place viewers figuratively at the Virgin's knees, with her head towering majestically above any observer looking at the picture from the church sanctuary.

The two little angels seated on the step below the Virgin's throne are especially significant as evidence of Masaccio's excited fascination with the inventive possibilities of Brunelleschi's new perspective. Each angel is shown strumming a mandolin, but notice that both instruments are depicted almost as if seen on end, foreshortened and slightly converging from either side and thus framing, even exalting by way of their focus, a special space leading to the Madonna's throne. The artist ingeniously realized that by so picturing them in the form of convergent diagonals they would give the impression that the whole fictive three-dimensional space was organized within a foreshortened lattice, from the front edge of the step on which the angels sit to the front edge of the raised step on which the enthroned Virgin is seated. Furthermore, if the angels were to stand up, their heads would fall precisely on Masaccio's horizon line crossing the Virgin's knees.

Masaccio seems to have intended his virtual space to be seen from a viewpoint related to the true height of human viewers before the picture, so that the mandolin-playing angels' feet do not rest on the implied floor below the throne, but only on another lower step, now invisible but presuming even more lower steps in order to give the impression that the whole ensemble sits high above the actual church floor.

This likelihood leads to another improvisation in Masaccio's expanding perspective imagination. It has frequently been noticed that two darkened streaks stain the tops of the two visible steps, entering from the left and projecting right about halfway across the panel. These markings are assumed to be painted shadows, as if cast from figures that must once have stood on that side beyond the present picture space. Some years ago the panel was removed from its frame, examined closely along its lateral and bottom edges, and, sure enough, evidence was revealed that it had once been cut down from what was originally a much wider and slightly longer picture. As has been mentioned, the lost side extensions showed paired saints standing conventionally, which on the left at least would account for the two mysterious shadows. Based on this new information, John Shearman attempted a reconstruction of what the Pisa Altar might once have looked like with all its missing parts restored (ill. 52).[8]

As we see in his published illustration, Shearman hypothesized the two shadows as cast, one by a single saint standing on the lower step left, the other by an additional angel genuflecting on the upper step at the same side. He then balanced these figures with a second angel and saint on the right, with the two other flanking saints standing on the same ground level as the seated, mandolin-playing angels in his presumed foreground.

Shearman invented these positions of the various figures without considering Masaccio's sophisticated perspective scheme. Illustration 53 shows my own reconsidered reconstruction, which is quite consistent both with the composition of his earlier 1422 triptych and his newly excited attention to Brunelleschian perspective, especially as evidenced by the peculiar shadow markings on the present painting's fictive steps.

Shearman has the two music-making angels in front of the Virgin as if their feet rested on a tiled floor, presumably extending fictively from the virtual pictorial space onto the actual floor on which the viewer stands. This cannot be, unless the diminutive size of the angels was intended to define the actual size of living observers looking at the picture from outside. As mentioned, if the angels were to stand up, their heads would precisely touch Masaccio's clearly defined horizon line at the Virgin's knees. We must then suppose that the artist intended the standing saints to be the size of giants and the enthroned Virgin perceived as even more elephantine in comparison (remember how Masaccio contained the traditional Trinitarian size of God the Father). Clearly Masaccio meant the angels to be less than human scale in

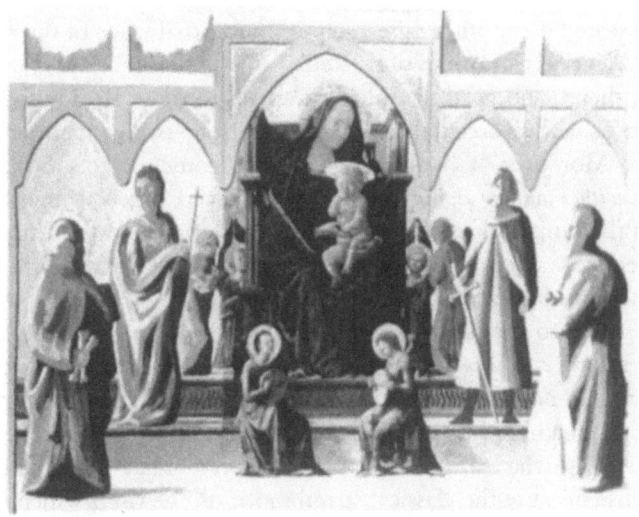

52. Reconstruction of Masaccio's Pisa Altarpiece by John Shearman, "Masaccio's Pisa Altarpiece: An Alternative Reconstruction," *Burlington Magazine* 108 (1966): 449–55. Photo courtesy of the Burlington Magazine.

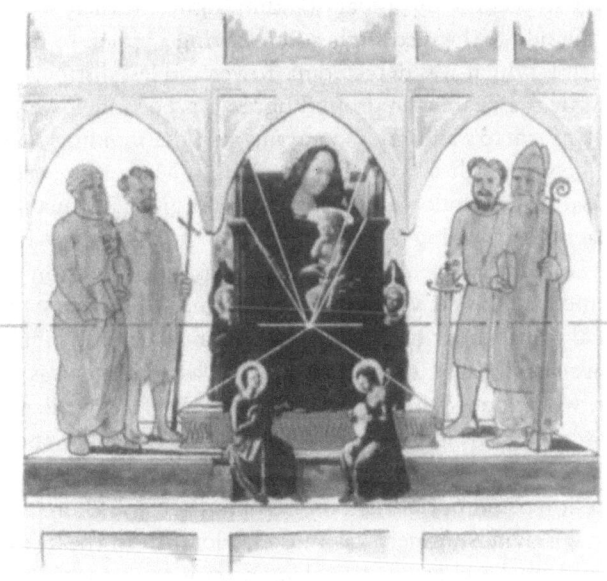

53. Reconstruction of Masaccio's Pisa Altarpiece by author.

size, and seated on another raised step in the virtual space of the picture, not at ground level as Shearman suggested.

Next, there is no need to add more angels to those already there because the painted shadows are easily shown to be cast by two conventionally standing saints. Moreover, Italian painters of such iconic subjects as the *Enthroned Madonna with Flanking Saints* rarely if ever placed angels in the same lateral spaces as the saints. Instead, angels always hover around the Virgin's throne and are entirely adjoined to her sacred space, just as Masaccio showed in his San Giovenale altarpiece.

Shearman also sought to extend the upper step clear across his expanded painting in order to account for the bit of painted shadow still visible upon it. However, I do not believe this was an extended step at all, but rather a narrow dais supporting the Virgin's throne, actually only a few inches wider on either side than what can still be seen in the present cut-down painting. This makes sense because the classical strigil motif on the riser, which spreads out in opposite directions from a central whorl, would never have been designed to run on indefinitely as Shearman has it, but would have been aesthetically applied only upon a short expanse of riser surface (in classical times, it often decorated the six foot or so sides of sarcophagi). In my reconstruction, this surface is but the short front of a square platform. The shadow on this started on the lower step as cast by a rear-standing figure. It then would have proceeded up the side of the next higher step leaving a bit left to fall visibly upon its top, which is why it was depicted narrower and less distinct that the wider shadow cast by the nearer foreground figure on the same lower step.

In both Shearman's and my reconstructions the standing saints are larger than the angels, yet still shorter than the Virgin Mary if she were to stand. I believe Masaccio still followed his medieval religious instincts here, perhaps even in reaction to having been criticized for making God the Father too small in his earlier *Trinity*. As mentioned, Masaccio's horizon and certification point indicate the viewing position of human observers standing outside on the actual church floor before the picture. In Masaccio's hieratic representation, however, because the entire painted assemblage was once upon a raised altar, worshippers standing in the sanctuary were privileged to observe this iconic scene not from the eye level of the tiny angels, of course, but from the presumed eye level of the standing saints; that is, if the latter were only able to step down and out of the picture onto the same church floor.[9]

In sum, as David Summers has mused, Florence in the 1410s and 1420s, even before Brunelleschi's decisive demonstrations in 1425, must have been abuzz with lively discussions among artists, as well as intellectuals, about what to make of the exciting new science of *perspectiva naturalis* and how painters and sculptors should take advantage of it.[10] Several, no doubt Ghiberti and Paolo Uccello as well as Donatello and Masaccio among them, were already zealously trying to find ways to adapt it. One of the earliest successful adaptations, again even before Brunelleschi, was made by one of the most surprising of Florentine painters, one who at the same time was presciently aware of the one great weakness in the new system—that its very geometry seemed to contradict a proper rendering of "spiritual space."

CHAPTER ELEVEN

More Masaccio, Masolino, and Even Fra Angelico

Masaccio's rapid assimilation of Brunelleschi's optical perspective grew ever more adjusted to his natural stylistic proclivities with each succeeding commission during the three last years of his short life. His next major work immediately after the *Trinity* was the decoration of the Brancacci Chapel (ill. 54) in the Church of Santa Maria del Carmine in Florence, which he undertook in partnership with his older friend, Masolino (ca. 1383–1447?).

This project was interrupted for a year by the interim Pisa Altarpiece commission, but that gave him even more time and experience in adapting perspective geometry to his innate sense of composition. In the Brancacci Chapel, for instance, we witness the first specific application of what I have called "horizon-line isocephaly." This optical phenomenon was only implicit in the *Trinity* and the Pisa Altar, just as it was in Brunelleschi's original demonstrations.

As already mentioned, the phenomenon has to do with the optical fact that when the viewer stands on level ground and observes a group of persons of equal height, just as the top edge of a quadrangular object at this level will appear as a straight line, so everyone's head will appear aligned at the same level, even though the feet of those persons further away seem to reach the "knees" of those nearer. Leon Battista Alberti finally put this rule into writing in his treatise on painting, *De Pictura*, of 1435:

> We see, in fact, in temples that the heads of moving men sway high at about the same height, while the feet of those placed farther away are perhaps corresponding to the knees of those who stand before [them].[1]

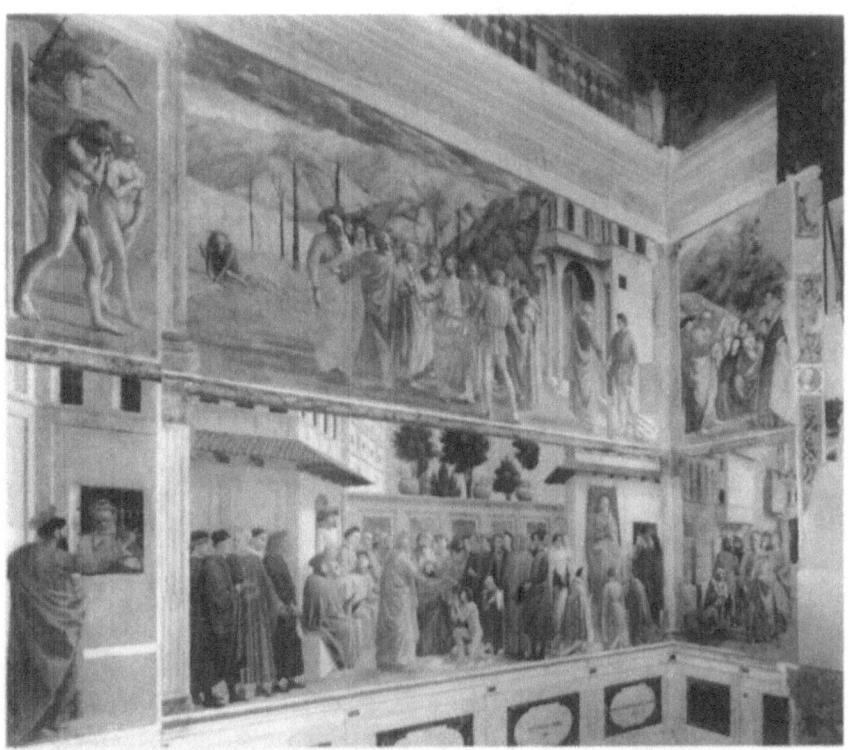

54. Left side of the Brancacci Chapel, Church of Santa Maria delle Carmine, Florence, 1425–27, ca. 1484. Fototeca Berenson, Villa I Tatti, Florence. Photo: Antonio Quattrone.

The clearest, and probably the earliest, artistic illustration of this principle is to be seen in the two upper-register frescoes on the left and right side walls respectively of the Brancacci Chapel, painted between the years 1425 and 1427, *Jesus and the Tribute Money*, by Masaccio (ill. 55), *Healing of the Cripple and Raising of Tabitha by SS. Peter and John*, by Masolino (ill. 56).

These scenes are depicted as if framed within a proscenium-like portico supported by classical columns on either side (as in illustration 54, the upper and lower frescoes on the left wall of the Chapel). In other words, viewers were supposed to imagine these holy stories as if taking place on architectural stages, just like the scenes of the life of Saint Francis inspired by miracle plays at Assisi. Furthermore, Masaccio and Masolino represented the human figures in their two frescoes as if all the heads were aligned along a continuous and common horizon, with certification points in the center of each

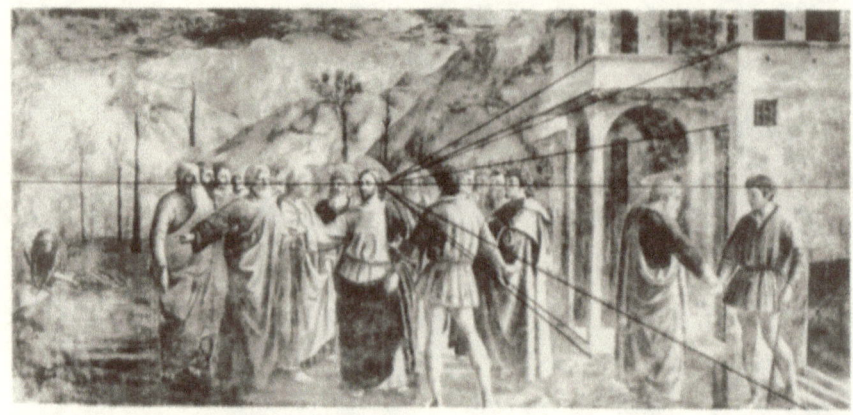

55. Perspective reconstruction by author of Masaccio's *Jesus and the Tribute Money*, Brancacci Chapel, Florence, 1425–27.

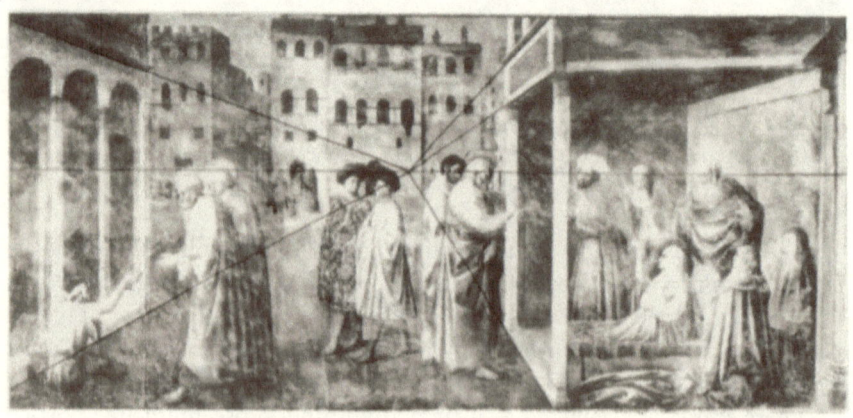

56. Perspective reconstruction by author of Masolino's *Healing of the Cripple and Raising of Tabitha by SS. Peter and John*, Brancacci Chapel, Florence, 1425.

scene respectively. The painted figures must then be understood as being seen by the artists figuratively standing on the same ground plane as if they—the artists—were suspended in the air. What both Masaccio and Masolino had now realized was that viewers, in order to perceive the same perspective illusion that Brunelleschi achieved in his initial experiments, did not necessarily need to stand physically in the same place from which the artist projected his picture via his own perpendicular axis. In fact, so long as the details within such pictures retain a coherent relationship to certification points and horizon line, viewers outside, no matter where they are standing as they look at such a picture, unconsciously imagine themselves vicariously identifying with the artist and seeing the scenes just as he saw them in his mind's eye from his personal vantage point.

Let us now attempt to reconstruct how these two painters applied their evolving, more idiosyncratic, perspective rules to the Brancacci Chapel commission. For one thing, they were already so familiar with the foreshortened grid as initial layout tool that they could compose it in their mind's eye and thus dispense with the actual drawing. Their most important adaptation of and improvisation on Brunelleschi's optical principles, on the other hand, was to take advantage of the horizon line as a compositional device, deploying it to unify aesthetically all the separate scenes in the Brancacci Chapel, four on each side wall and four more on the end altar wall. (Before the present architectural altar was placed there in the eighteenth century, at least one more large scene had been painted in the center of this wall but had subsequently been destroyed).

Masaccio and Masolino apparently divided their work, with the left side of the Chapel to Masaccio and the right to Masolino (but merging their positions when painting the end wall). Together again, they decided that a common "isocephalic" horizon line should unite all their frescoes on both sides in the upper registers; that is, all the figures painted standing in these scenes should have their heads at the same level. Furthermore, the certification points on the horizons of the frescoes on the long walls should be located precisely on the center of each (ill. 57). Although Masaccio and Masolino certainly initiated this schema and no doubt intended to apply it to the lower walls as well, they were both suddenly called away to other commissions (and to Masaccio's untimely death in 1428). Not until some sixty years later was the commission finally completed, by Fillipino Lippi (1457/8–1504) whose additions, thankfully, adhered precisely to the earlier masters' perspective layout.

On the upper side-wall register to the left, Masaccio painted his famous narrative (ill. 55), while on the adjacent upper-register right wall, Masolino

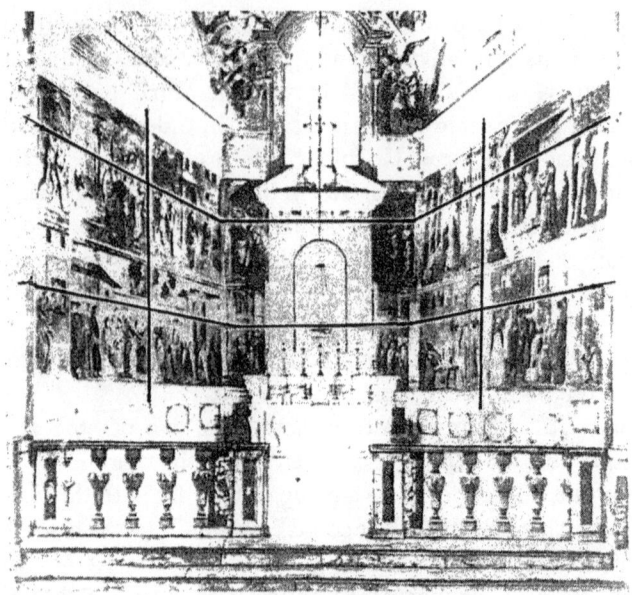

57. Reconstruction by author showing how the artists employed perspective to compose and unify the scenes on all three walls of the Brancacci Chapel.

frescoed his signature masterpiece (ill. 56). It is worth comparing how these two close friends, often so similar in painting style, differed significantly in their individual interpretations of Brunelleschi's new perspective rules.

Although both agreed that single horizon lines should unify all the scenes in upper and lower registers, and that single certification points should be placed in the center of these lines on the side-wall frescoes, each artist harbored a different conception of how such optical principles should affect the overall pictorial organization. With Masaccio, for instance, the certification point is much more obvious than the horizon line, while with Masolino, just the opposite is true. Masaccio emphasized the certification point by locating it directly upon the head of Jesus, underscoring that Christ was the principal figure both iconically and optically in the whole composition,[2] while Masolino's certification point lies in an innocuous empty space unassociated with the iconic narrative (the hole in the fresco where the artist placed a nail in order to draw the architectural orthogonals can still be traced to this day, just to the right of the two strolling dandies in the middle of the pictured piazza).

Masolino seemed to be especially excited by the dramatic, geometric effect of his horizon, aligning all the heads, even of tiny distant figures in the
far background, as if strung on a stretched clothesline. In fact, he grew almost obsessed with such an effect in his later work during the 1430s, even as
his style, after Masaccio's death, tended to revert to a flatter, less illusionistic
manner. This is evident in a series of frescoes he executed for Cardinal
Branda Castiglione in the Roman Church of San Clemente, and in the Cardinal's home town, Castiglione Olona in northern Italy, particularly the
scene of *The Feast of Herod* painted in the local baptistery (ill. 58).

Once again the artist positioned his centric certification point in a blank
space beside the heads of a couple of extraneous figures, between the two
important narrative events on either side in the foreground (the *Feast* to the
left, and *Salome Presents the Head of John the Baptist* to the right). However,
what truly excited the artist was the stark linear effect of the horizon line,
not only stringing together the heads of the foreground figures but the collective capitals of all the columns of an extraordinary arcade to the right,

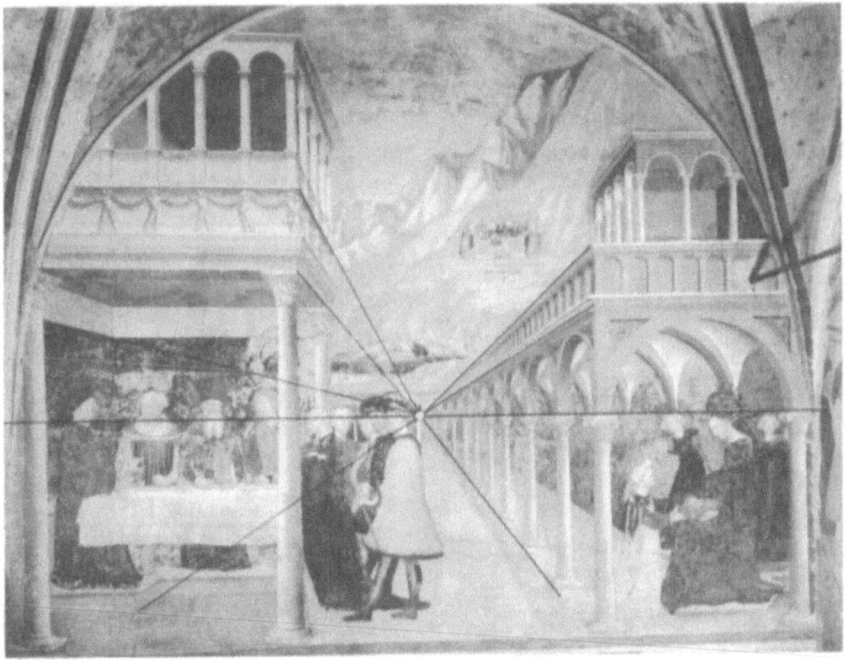

58. Perspective reconstruction by author of Masolino's *Feast of Herod*, Baptistery, Castiglione
Olona, 1435.

whose front face in the picture is incongruously flattened to its perpendicular side. Oddly, Masolino seemed little interested in the illusion of depth, but quite idiosyncratically applied Brunelleschi's perspective rules solely to create novel but still rather flat Gothic-like surface patterns.

In his 1435/6 treatise Alberti also described a shortcut for proving the geometrical correctness of the optically foreshortened grid. To demonstrate the accuracy of his method, he inscribed a line from corner to corner of his foreshortened gridded pavement. If it simultaneously intersected the corners of each of the squares through which it passed, the diminishing transverse spaces between them had been correctly drawn. Illustration 59 shows how this is so.

Whether or not Brunelleschi would already have discovered this "proof" on his own, and then perhaps to have shown it to his young friend Masaccio, the artist of the Brancacci Chapel frescoes certainly understood the basic geometric rationale. In his scene *St. Peter Healing with His Shadow* on the lower left end wall (ill. 60), he left a revealing palimpsest, the incision marks in the detail of the receding balcony above the heads of the sick and crippled bystanders being cured by Peter's passing shadow (ill. 61).

What the painter was trying to depict in this detail were the three equally separated supporting struts under the overhanging balcony above the heads of the standing figures. The balcony, of course, was to appear as receding diagonally into the distance behind St. Peter as he advances in the foreground. Thus, the spaces between the balcony struts must likewise be made to diminish slightly, the nearer seeming larger than the further, just like the

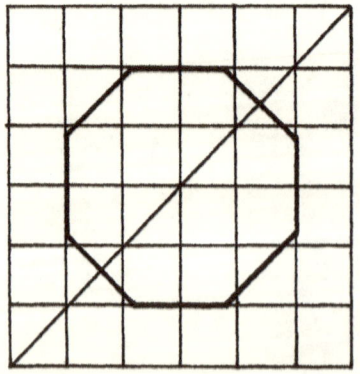
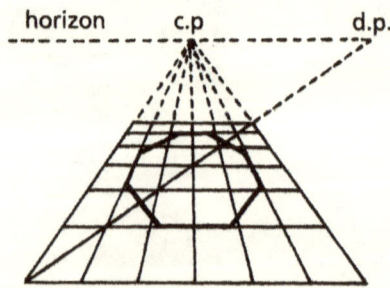

59. Alberti's perspective proof. Drawing by author.

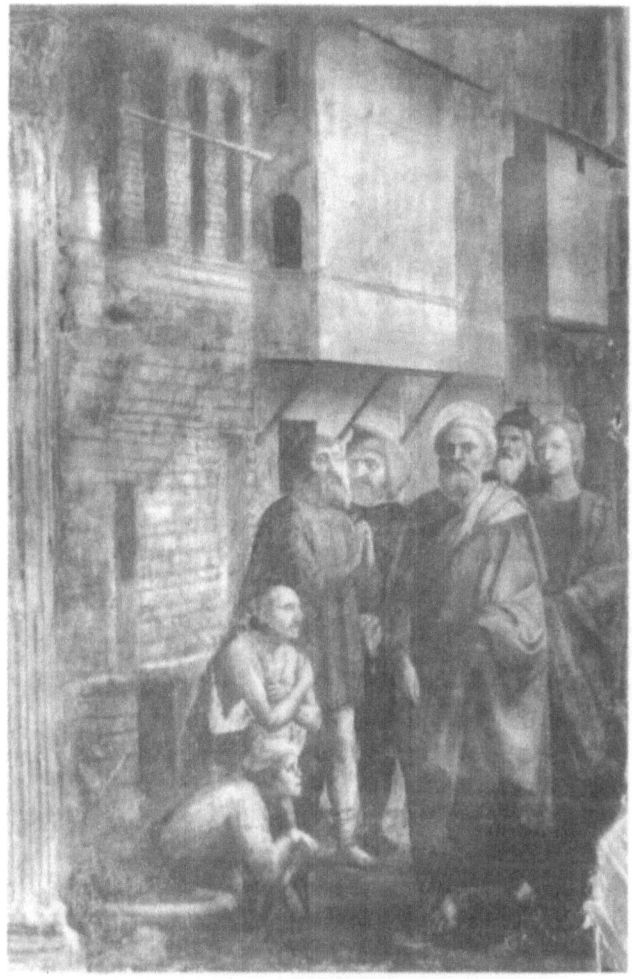

60. Masaccio, *St. Peter Healing with His Shadow*, Brancacci Chapel, Florence, 1425. Photo: Gabinetto Fotografico del Polo Museale Fiorentino.

transverse spaces of a foreshortened gridded floor. We are able to realize exactly how the artist measured this from the still visible incised lines, as in my reconstruction, illustration 62.

Masaccio first marked out a line on the painted wall beneath his fictive balcony where the presumed evenly spaced supporting struts should attach to the wall. This line, along with the parallel bottom edge of the balcony

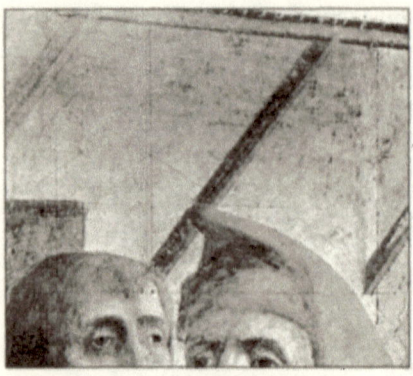

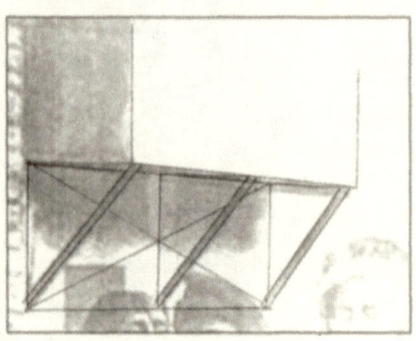

61. Detail from illustration 60 showing incised lines shaping the perspective recession of the wall behind St. Peter.

62. Enhanced drawing by author of illustration 61.

against the wall, should seem to converge slightly, framing a trapezoid, which, of course, is what a receding rectangle looks like in perspective. However, in order to position correctly the center strut, especially where it attached to the wall, Masaccio needed to locate the center of the foreshortened rectangular space on the wall, which he did precisely by the same means that Alberti later prescribed, but which in Masaccio's case was no doubt another derivation from the old "bifocal construction"; that is, by extending lines from both corners across to their diagonally opposite corners, and marking (with a distinct "X") where they intersected. This automatically established the illusionistic "center" of the foreshortened rectangle, and thus established exactly where to place that middle support according to Brunelleschi's new *perspectiva artificialis*.

Finally, the most surprising (but, in the context of this book, not so surprising) Florentine painter during the 1420s caught up in the popular fascination with the spiritual and moral implications of *perspective naturalis* was none other than Fra Angelico (ca. 1400–1455). His true given name was Guido di Pietro, changed to Fra Giovanni when he joined the Dominican order. Shortly after 1418 he moved into the new convent in Fiesole where he became associated with Fra Antonino. For the next twenty years, the two were such close colleagues that in 1439, when Fra Antonino moved to the Convent of San Marco and soon became its prior, he personally allowed his old friend to direct the mural painting of all the monks' cells there. This great master project has since been acclaimed as one of the most beautiful set of frescoes in the entire Renaissance. In recognition, the Italians awarded Fra

Giovanni the pious name he has been known by ever since, Beato Angelico, "Blessed Angelic."

For years, however, Fra Angelico has suffered because of his name, especially during the nineteenth century, which regarded him as a charmingly *retardataire* painter still clinging to the old stylisms of the Gothic Middle Ages. In truth, his often saccharine-looking, iconic Madonna-with-Child paintings do seem to ooze pure Victorian sentiment. Even his then presumed birth date, 1387, positioned him as an elder conservative eschewing the advances being made by the younger, more fast-lane Florentine painters like Masaccio. Only until his "old age" after the 1430s was Fra Angelico's style believed to have caught up to that of his contemporaries, allowing him to be included among the truly progressive painters of the Renaissance. Although several of his early paintings do seem to exhibit a certain perspective familiarity, these were always thought to have been done in the very late 1420s, or in the mid to late 1430s, when they could be better explained as merely reflecting Masaccio's established art, or the au courant theories of Leon Battista Alberti.

But in 1954 new documents were discovered indicating that Fra Angelico was actually born later, nearer to 1400, making him a close contemporary of Masaccio.[3] Furthermore, in an exhibition of the artist's early paintings at the Metropolitan Museum of Art, New York,[4] a number of lesser-known, small paintings with surprisingly unexpected perspective hints were included that, it now turns out, must be documented as having been done well before 1430, eliminating any possible influence from Alberti, and some even a few years prior to 1425, before Brunelleschi's and Masaccio's public display of perspective rules. The only way to explain this unexpected precocity was to admit that Fra Angelico was likewise among those intellectuals and artists thinking about the new science of geometric "*prespettiva*" and how to apply it to art, almost as early as Brunelleschi and even earlier than Masaccio.

One painting in the Met exhibition was particularly revealing in this respect, a *Madonna and Child Enthroned* from the private collection of Barbara Piasecka Johnson (ill. 63).[5] This extraordinary work has remained largely unknown to the art historical community. It was only published for the first time in 1963, when it was quickly accepted by most scholars as a fully autograph work by Fra Angelico. What made this attribution so significant was that the painting, surely dated no later than 1422, bears striking similarity to Masaccio's 1426 Pisa *Madonna* (ill. 51). Could it be possible that the latter was influenced by the former? This was tentatively suggested by Anneke De Vries, writing in the 2005 Metropolitan Museum catalog edited by Laurence Kanter and Pia Palladino.[6] Nevertheless, to avoid such a troubling analogy,

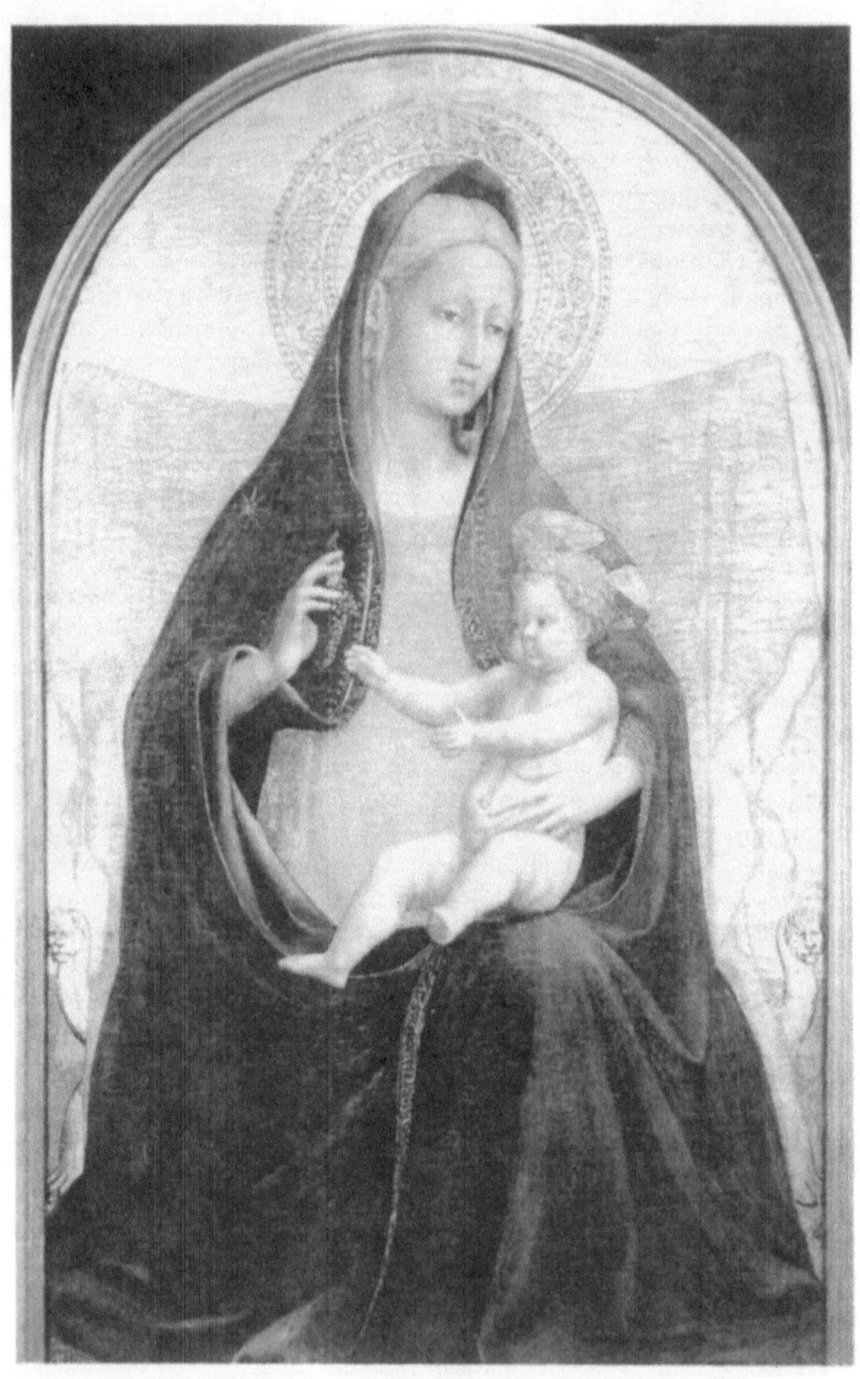

63. Fra Angelico, *Madonna and Child Enthroned*, 1422. Property of the Barbara Piasecka Johnson Collection Foundation.

other art historians have reverted to conventional wisdom and opined instead that the painting should really not be compared to Masaccio, but only to Fra Angelico's earlier antecedents, and particularly to his preceding San Domenico altarpiece, the artist's first major contribution to this genre, which he painted for the high altar of the new convent church in Fiesole around 1420, shortly after he joined the Dominican brotherhood there (ill. 64).[7]

The well-known central panel of this once multipaneled altarpiece is still in situ in the San Domenico church (although now relegated to a side chapel). It again shows the *Madonna and Christ Child Enthroned* surrounded by a cluster of tiny angels and four standing saints. Unfortunately, much of Fra Angelico's original surface has been repainted. A fictive architectural backdrop replaced the original gold background, and even the outer frame was "modernized" in accord with sixteenth-century tastes. In any case, the central arrangement of

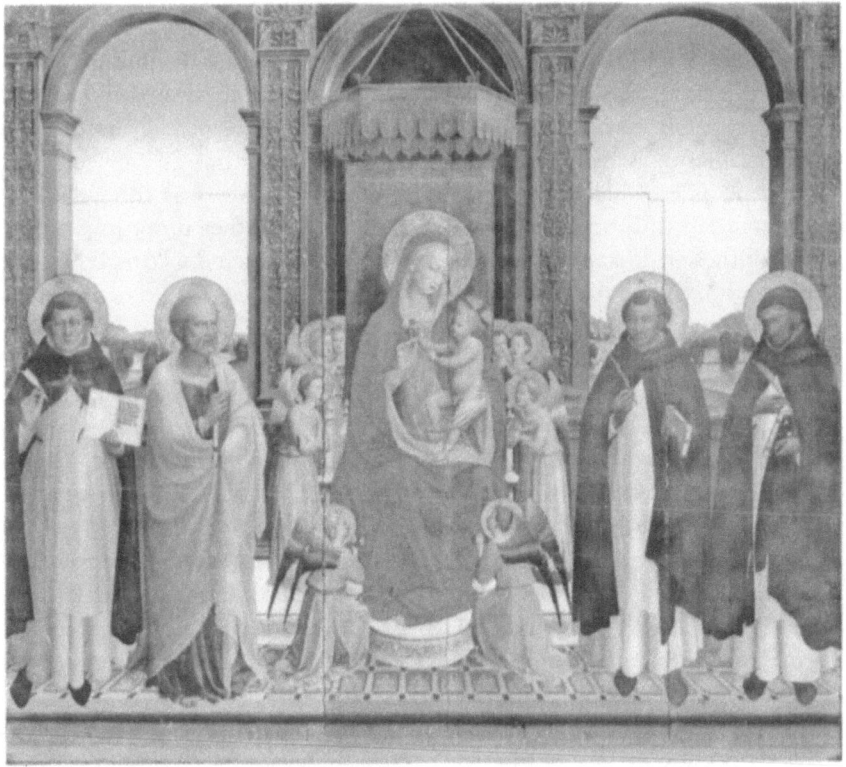

64. Fra Angelico, *Madonna and Child Enthroned*, Church of San Domenico, Fiesole, ca. 1420. Photo from author's collection.

figures remains Fra Angelico, and one critical detail especially points to this painting, as well as the above mentioned Piasecka *Madonna and Child*, quite distinctly away from its fourteenth-century forebears and directly toward Masaccio.

Notice the halo over the head of Baby Jesus in both paintings. It is clearly in the shape of a tilted ellipse, similar to the elliptical halo attached to the head of the same Baby Jesus in Masaccio's Pisa Altar. We have here a very revolutionary innovation surprisingly overlooked by modern critics. Fra Angelico seems indeed to have been the first painter in all of Western Christian art to have imagined the heretofore mystical halo as a geometric object in three-dimensional space subject to the perspective point of view of the earthly observer in front. He set the halo no longer behind the Infant's head, but as if affixed to the top of it, so that the circular form of the halo must be seen distorted into various elliptical shapes depending on the turn of the Infant's face.

Medieval Christian paintings always represented the halo as a flat aureole behind the heads of saintly figures. This often caused some unhappy compromises among the more inventive artists, particularly as they became ever more interested in having their imagery imitate visible nature. For instance, Giotto, when he wanted to depict holy figures as viewed from behind in some of his Arena Chapel frescoes (ca. 1305), was compelled to have them face incongruously into their halos.[8] Another unsettling problem with these halo anomalies, however, occurred when the Baby Jesus was shown nestled high in his mother's arms. How then to place the Child's halo? If directly behind his head as was traditional, then the halo must block much of the Madonna's own face. This problem was usually avoided by having the Baby's halo tucked half hidden behind the Madonna's shoulder, still an incongruous solution, but at least preserving the Madonna's dignity. Both Fra Angelico and Masaccio no less often resorted to hiding Baby Jesus's halo in this way, as for instance in Masaccio's San Giovenale Altarpiece of 1422 (ill. 50).

We can well imagine Fra Angelico discussing the matter with Fra Antonino at some point. Recall that Fra Antonino had lamented in one of his sermons that some artists were led to create "oddities . . . monstrous in the nature of things," when they represented certain arcane Christian mysteries.[9] Perhaps indeed Fra Antonino suggested to his friend that he repair the troubling halo oddity by picturing it *in figuram*, "into a figure," as he often advised, employing the accusative case in order to emphasize how one should translate mental forms of holy images into visualizable "spiritual geometry."

Fra Angelico actually experimented with such elliptical haloes in a number of paintings during the early 1420s, and maybe a little before, but he never stopped to work out the precise measure of its perspective foreshortening as Masaccio clearly did in his 1426 Pisa panel.[10] Subsequently, and probably due more to the latter's refinement than Fra Angelico's tentative and largely unnoticed examples, the foreshortened halo, attached almost as if screwed to the top of the head, became a commonplace convention everywhere in Renaissance-style art. Fra Angelico, on the other hand, seems to have abandoned this idea in most of his later paintings, returning to favor the traditional flat gilded disk.[11]

But let us review again Fra Angelico's San Domenico Altarpiece (ill. 64), its date now retracted from ca. 1428 to ca. 1419–20, painted just after the young novitiate joined the newly established Dominican convent in Fiesole. As mentioned, the painting displays the Madonna and Child surrounded by eight tiny angels closely gathered about her throne, six on the raised dais on which the throne sets, and two on the squared pavement below directly flanking the seated Virgin. This central composition is in turn flanked by four saints, arranged two on either side and all standing on the same squared pavement.

Fra Angelico's San Domenico Altarpiece indeed remarkably resembles my previous chapter reconstruction of Masaccio's original Pisa Altarpiece (ill. 53). When I first began to work on the Masaccio ensemble many years ago, I made no attempt to relate it to Fra Angelico, other than believing like most art historians that the latter's painting was done several years later than 1426 and thus must be derivative of Masaccio. I now believe, based on the new dating and the similar apparent correspondence of the Piasecka Madonna, that Masaccio not only admired Fra Angelico's halo innovation, adopting and perfecting it in all his painting after 1425, but may also have borrowed Fra Angelico's entire San Domenico composition, which seems already to have influenced his 1422 altarpiece.

For instance, Fra Angelico's angels are similarly smaller than human size and clustered closely around the Madonna with none intruding into the spaces occupied by the larger saints. The two angels before the Madonna's throne, even though holding no musical instruments, do likewise seem to have inspired Masaccio. Although Fra Angelico was not yet incorporating precise optical principles concerning certification point and horizon line, he did, just as in the Pisa panel, have the two foreground angels frame an empty space implying fictive depth before the Madonna's majestic throne, emphasizing the special distance viewers should respectfully maintain as they confront the divine presence. Note too that this throne is elevated on a raised dais with only short extensions at either side. Below the dais, all the saints stand on a common level, just as I have contended regarding Masaccio's panel. This lower level is also shown with a step going down in front of it, again like the Pisa panel, and, as I have further claimed of the latter, implying that the saints are standing above the main floor of the fictive structure in which this scene is taking place. They were thus intended to represent the same size as ordinary human viewers of the picture.

And finally one more of Fra Angelico's prescient, now predated paintings, the *Deposition from the Cross* in the Museo San Marco, Florence (ill. 65). This large altarpiece once stood in the Church of Santa Trinita in Florence, having been commissioned by the rich banker Palla Strozzi in 1423 as a companion piece to Gentile da Fabriano's *Adoration of the Magi* (ill. 75).[12]

The original painter was supposed to have been Lorenzo Monaco (1370/72–1423), Fra Angelico's early mentor and far the more *retardataire* stylist. The artist died prematurely, however, and sometime around 1430–31 the work was reassigned to Fra Angelico. We recall that Fra Antonino had spoken out against the very same "vanities" depicted so profusely in Gentile's companion panel. One can well imagine that the stern Dominican then recommended Fra Angelico to complete this second Strozzi commission as a sober antidote to Gentile's raucous *Magi*. In any case, Fra Angelico's painting, as eventually finished by 1432, is quite representative of his emerging "classical" style.[13] Indeed, the limp figure of Jesus being lowered from the cross is reminiscent of ancient depictions of the dead Meleager, hero of the Trojan War, as Leon Battista Alberti described him a few years later: "all the members hang down, the hands, the fingers, the neck: all fall down limp. . . . all contribute to express the death of the body."[14]

No doubt Alberti was already familiar with Fra Angelico's *Deposition*, and he may well have had this melancholy image on his mind as the ideal model for depicting the tragic remains of ancient warriors on the classical battlefields of his *historiae*.

Furthermore, Fra Angelico's *Deposition* reveals another detail that would have attracted Alberti. I have inscribed a straight line horizontally through my posted photograph (ill. 65), which aligns more or less with the heads of nearly all the standing figures in the painting, particularly that of Joseph of Arimathaea, who is supporting Jesus's body in the exact "certification point" center of the composition. My superimposed line clearly indicates that the artist was already aware of the "horizon-line isocephaly" principle, no doubt reinforced after having studied Masaccio's *Tribute Money* in the Brancacci Chapel.

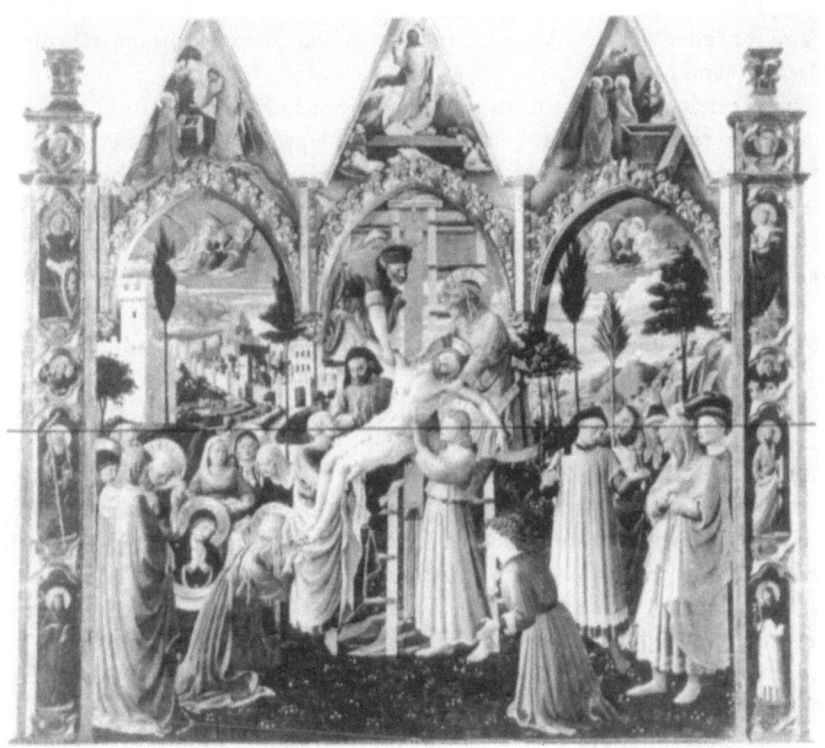

65. Perspective reconstruction by author of Fra Angelico's *Deposition from the Cross*, 1432. Museo San Marco, Florence.

Look now more closely at the small painted figures of six saints, three on the front surface of each of the two tall finials at the sides of the frame. Notice that each saint stands on a circular pedestal. Note further that the pedestals under the lower four saints on either side are tilted in such a way that implies that they are being viewed from an eye-point above, while the pedestals under the upper two saints on either side are tilted in such a way as to imply that they are being seen from below. My reconstructed horizon line runs precisely between these upper two and lower four pedestals, more proof if any is needed that Fra Angelico was just as cognizant of this optical rule as Masaccio and Masolino.

Nevertheless, in spite of his precociousness, Fra Angelico from the very beginning recognized a problem regarding the projection of sacred imagery in geometric perspective. How should he differentiate between what one sees with corporeal eyes and what must be depicted as if seen with spiritual eyes, just as Fra Antonino advocated? Recent studies by Patricia Rubin, Megan Holmes, and Marcia Hall have shown that the Dominican priest-painter was often troubled in trying to indicate this separation, sometimes applying two certification points, one low for picturing earthly space, the other high to signify the heavenly realm (for instance his *Coronation of the Virgin* now in the Louvre).[15] He would even revert to adding traditional empirical conventions mixed in with his up-to-date perspective (for example, the details of hell scenes in his San Marco *Last Judgment*). Like hesitant Saint Thomas in Verrocchio's Orsanmichele sculpture (ill. 8), Fra Angelico was willing to look *at* the utterly sacred, but unwilling to actually extend himself *into* it; that is, to define it geometrically as an equal space contiguous with his own. He seemed to believe that geometry, by its very Greek-derived name, "earth measure," was unworthy, especially when the rules were followed to the letter, to reveal what is heavenly and therefore unmeasurable by mundane mathematics.[16]

This apparent equivocation in Fra Angelico's application of the new perspective has led many earlier critics to assume that the artist never learned the rules fully, or that these ambiguous passages in his several paintings were inserted by less able assistants.[17] We should now realize that whenever such inaccurate perspective instances occur in Fra Angelico's art, they were most likely intended by the master himself and hardly due to naïveté.

CHAPTER TWELVE

Alberti's Method

Leon Battista Alberti, the scion of a noble Florentine family exiled in Genoa for political reasons, yearned all during his youth to return to the city of his heritage. Meanwhile, he became highly educated in the classical Latin humanist tradition, and for several years served as secretary in the papal chancery in Rome. Only in 1428, with a favorable change of government, was he able to return for the first time to Florence. Indeed he was so impressed and excited by the extraordinary flourishing of art and architecture during those fecund early Renaissance years that he quickly, between 1435 and 1436, wrote his famous *Treatise on Painting* in two editions, first probably in the Tuscan vernacular (eventually printed in Italian as *Della pittura*), and then in Latin (*De Pictura*), which he reedited in several revisions over the next thirty-some years. Each version, however, consisted of three books, the first describing the geometric rudiments of painting, the second on composition, drawing, and coloring, and third on the ethics and morals of the artist.

Scholars remain in some disagreement as to which language edition came first, but the strongest evidence to date suggests that the vernacular preceded the Latin.[1] The vernacular edition indeed was dedicated to no less than Filippo Brunelleschi, along with honorable mentions to Masaccio, Donatello, and Luca della Robbia. However, only three manuscripts of this text are extant today, indicating both its limited audience and localized circulation, while there are twenty copies of the Latin text in various European libraries, indicating that this was the language in which the author chose to spread his ideas to the widest possible public (at a time when all educated Western Europeans both read and spoke Latin). Moreover, it was the finalized Latin text

117

that ultimately was printed in 1540 and thus was the most accessible and influential version from the sixteenth century on.[2]

As mentioned, Alberti divided his text into three short books. The first, and only one to be discussed in this chapter, established the geometric foundation of painting, beginning with basic Euclidian definitions of point, line, and plane, and then from classical optics, the concept of visual rays and visual pyramid, all leading up to his perspective method based on the intersection (*intercisio*) through the visual pyramid. As I have also already insisted, the geometry and optics of Alberti's perspective pretty much derived from Brunelleschi's 1425 mirror demonstrations. Nonetheless, the author did reorganize what may have been a rather confusing explanation by Brunelleschi *viva voce* into a simplified step-by-step procedure that any literate painter could follow, although, curiously, without benefit of visual aids. In fact, in none of the nearly two dozen known manuscripts in both languages surviving from the fifteenth century are there any illustrations, leaving modern scholars to reconstruct from Alberti's verbal texts what seems to have been in his own mind's eye.

Here follows Alberti's long description. Seven of my own diagrams are interspersed among the translated paragraphs below, each illustrating a sequential step:

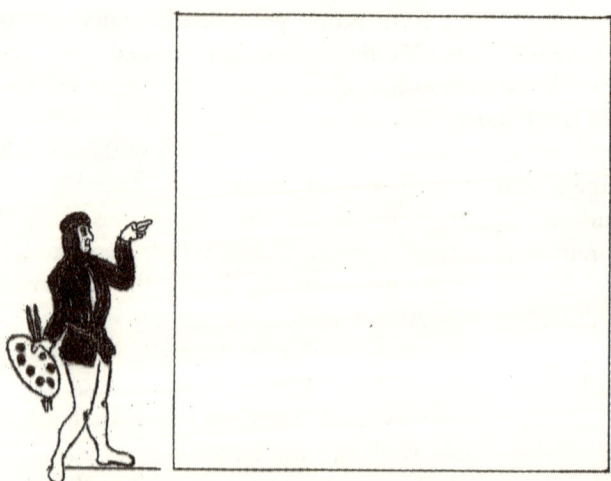

66. Step-by-step reconstructions by author of Alberti's perspective method.

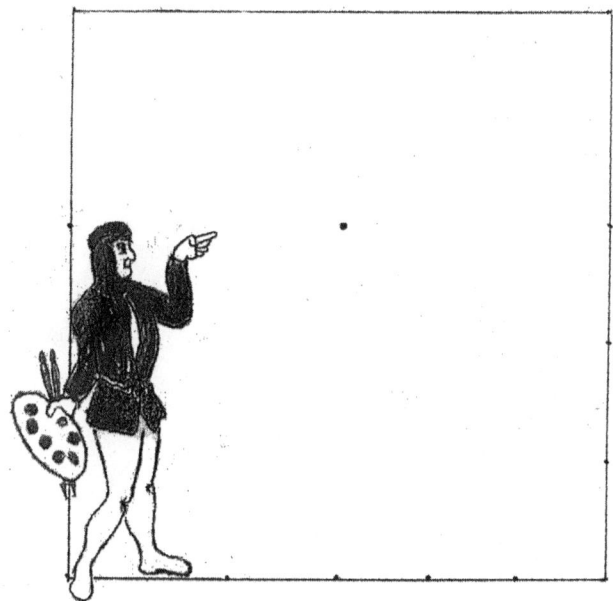

67. Step-by-step reconstructions by author of Alberti's
perspective method.

I will say what I myself do when I paint. First, I trace a large quadrangle, as I wish, with right angles on the surface to be painted, in which place it certainly functions for me as an open window through which the *historia* is observed, and there I determine how large I want men in the painting to be (ill. 66).

And I divide the height of this very man into three parts that for me are certainly proportional to the measure that the people call *braccio*. That [measure] of three *braccia* as it results from the symmetry of the limbs of a man, is commonly the height of the human body. According to this measure then I divide the base line (*iacentem infirmam*) of the drawn quadrangle into as many parts of this kind that the [line] contains. Moreover for me this very base line of the quadrangle is certainly proportional to the nearest transverse and equidistant quantity seen on the pavement (*pavimentum*).

After these things, I place only one point inside the quadrangle; in that place let there be the [point of] sight. For me, that point as it occupies the place itself toward which the centric ray strikes, let it be called, therefore, the centric point (ill. 67).

The appropriate position of this centric point is not to be higher [above] the base line than the height of that man to be painted. By this agreement, both the

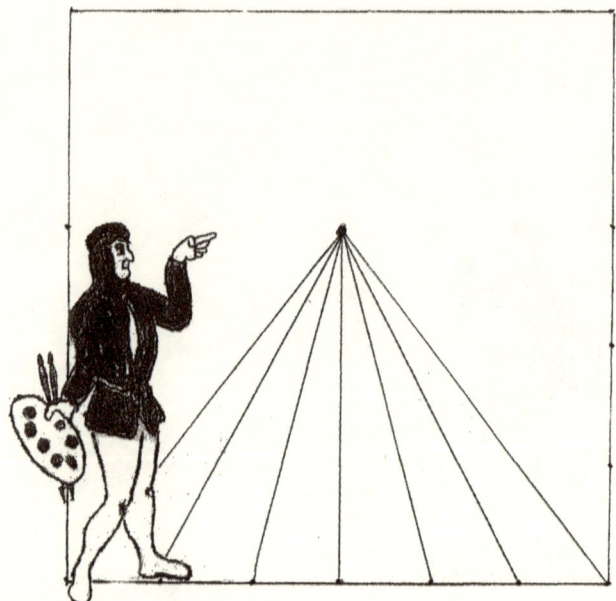

68. Step-by-step reconstructions by author of Alberti's perspective method.

viewers and the painted things appear to be on a uniform plane. Having placed the centric point, I draw straight lines from that very centric point to the single subdivisions of the base line. Which lines certainly show me how the transverse quantities, wanting to advance by interval, are drawn together in vision up to an almost infinite distance . . . (ill. 68).

But in [drawing] transverse quantities I observe this method. I have a small surface on which I trace one straight line alone which I subdivide into those parts according to which the base line of the quadrangle has been divided. Hence I place up from that line a single point as high as the centric point is in the quadrangle [and as] distant from the subdivided base line. From this point I trace individual lines to the individual subdivisions of this very same line (ill. 69).

I then establish how much I want the distance between the viewer's eye and the painting to be, and having fixed there the position of the intersection, I effect by means of a perpendicular line, as the mathematicians say, the intersection of all the lines that this perpendicular will have come between. . . .

This perpendicular line, then, will give me, through its subdivisions, the limits of every distance that must occur amongst the equidistant transverse lines

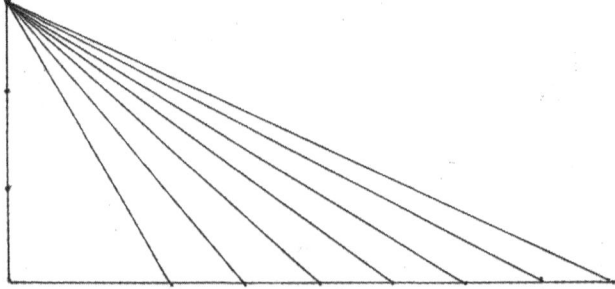

69. Step-by-step reconstructions by author of Alberti's perspective method.

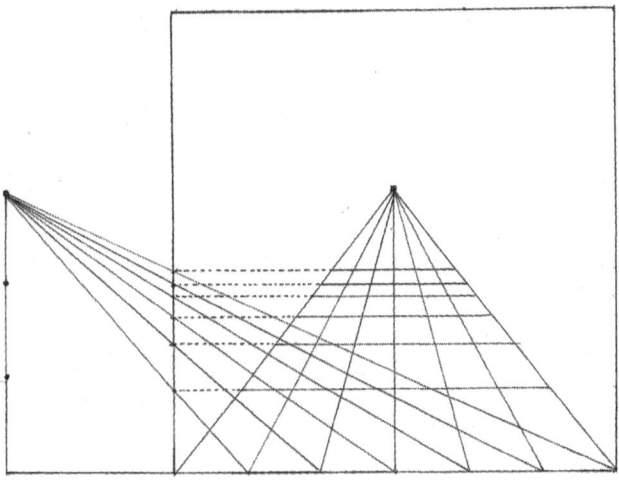

70. Step-by-step reconstructions by author of Alberti's perspective method.

of the pavement. In this manner, I have represented all the parallels of the pavement (ill. 70).

An indication of how these [lines] have been drawn in a correct way will appear when one single straight line extended becomes the diagonal of the connected quadrangles in the depicted pavement.

The diagonal of a quadrangle is certainly, for mathematicians, a certain straight line carried from one angle to its opposite; which divides the quadrangle into two parts so as to make two triangles from the quadrangle. After having done these things to perfection, I trace in a similar way one single transverse line

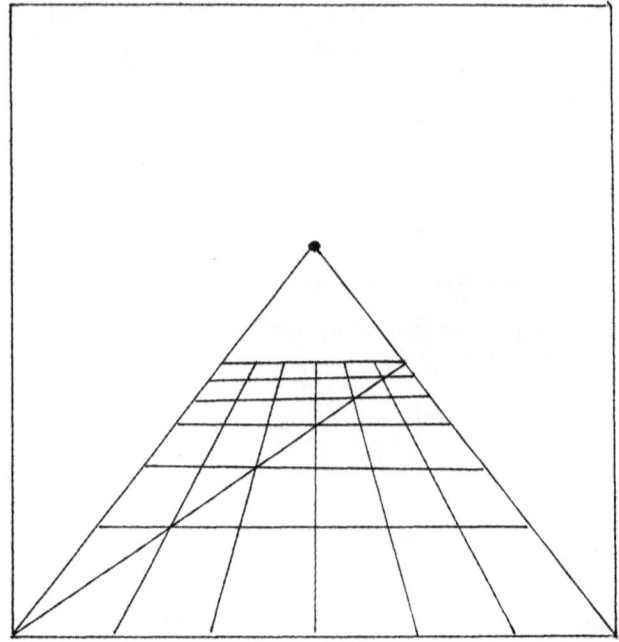

71. Step-by-step reconstructions by author of Alberti's perspective method.

parallel to the remaining ones below, and which intersects the two vertical sides of the large square and passes through the centric point. This line indeed constitutes for me the termination, or rather the limit [above which] nothing higher than the viewer's eye exceeds. And because it passes through the centric point, let it therefore be called the centric [horizon] line. It follows that depicted men who would be standing on a parallel further away, however much smaller [they may seem] than those placed on the nearer parallel, nevertheless [are] not smaller with respect to the others, but [only] further away. It is evident that this phenomenon is clearly revealed by nature herself. We see in fact in temples that the heads of moving men sway high at about the same height, while the feet of those placed farther away are perhaps corresponding to the knees of those anterior (ill. 72).[3]

Although Alberti made no direct reference to the ancient discipline of optics anywhere in his treatise, and never mentioned the Latin word *perspectiva* or its Italian equivalent in either edition, there can be no doubt that he understood that his readers would already be familiar with that recently revived and currently popular science and thus recognize its sacred authority as

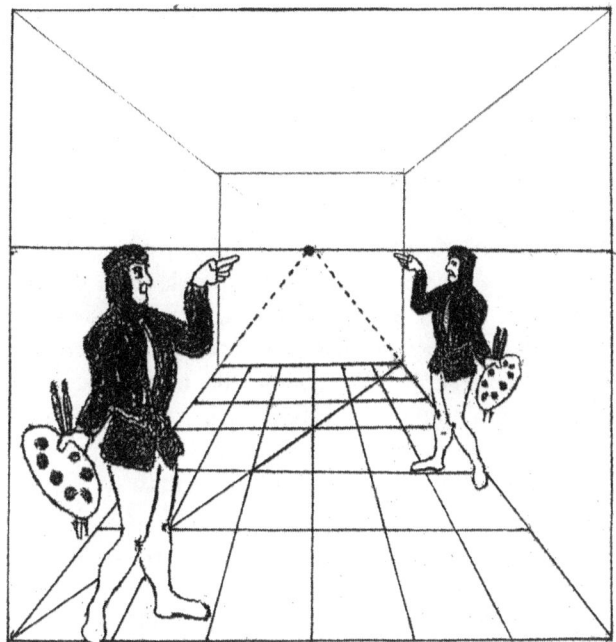

72. Step-by-step reconstructions by author of Alberti's
perspective method.

the theoretical basis of his method. Nevertheless, although Alberti began his
treatise by rehearsing the basic Euclidian tenets of point, line, and circum-
scribed plane as abstract mathematical concepts, he did redefine them as
tangible physical forms such as artists would draw with brush on wall or
panel. He even substituted new objective words such as *signum*, or "sign," for
a point, and *fimbria*, or "brim," to describe the outline of a depicted plane.

It's noteworthy, as shown in illustration 72, that Alberti assumed, without acknowledging, Brunelleschi's discovery of the optical coincidence of
certification point and horizon. However, in his continuing efforts to avoid
abstruse language (and credit to Brunelleschi), he termed the principle simply as "centric." However, he did seem to be unclear as to whether his
centric point should connect with the eye level of his generic viewer or
with the top of his head. Whichever, Alberti would simply insert a vertical
as high as his generic man up from his baseline, and from the tip of this all
sight lines were then to be directed toward the divisions on the pavement
(ill. 69). He also did not indicate whether the "intersection" of all these
sight lines was to be at the perpendicular edge of his picture quadrangle,

the easiest way of drawing it as I have done in my illustration 70, or at the central orthogonal of the pavement itself, which requires more careful alignment but is truly the more accurate way of establishing "how much I want the distance between the viewer's eye and the painting to be." No doubt Alberti had already realized that it didn't really matter where the viewer should stand in order to sense the illusion of perspective in a picture, only that the geometry of its optical projection be correct from any distance the artist chose, near to the picture if the pavement was to be steep or more removed if the pavement was to be more level.[4] Like other modern scholars, I remain uncertain as to just what Alberti intended.

Indeed, his directions seem to have been recorded differently in several of the manuscript copies of De Pictura. Some, like the present translation from the 1540 printed edition, imply that he first worked out the diminution of the transverse spaces on a separate piece of paper, and then transferred the results across the orthogonals of his pavement. Others seem to say that everything is to be done in the same drawing.[5] In any case, his added proof provided yet another shortcut. As my illustration 71 shows, the artist could even construct an optically correct foreshortened floor without any distance point construction at all. Nevertheless, whatever Alberti's actual method, his perspective layout made it possible to draw objects in proper foreshortened size anywhere in fictive pictorial distance automatically according to which transverse row of squares the seen object is set upon. Alberti even described how to inscribe a circle within the gridded trapezoid so that it would appear as a correctly shaped ellipse in perspective, and also how to project the vertical sides of a building directly up from the squared pavement, techniques as I have argued were already in practice by a number of Florentine artists since Brunelleschi's 1425 demonstrations.

Around 1445, the Veronese painter Pisanello (ca. 1395– ca. 1455/6) made a practice drawing, perhaps the first to illustrate Alberti's directions directly from one of the manuscripts now circulating outside Florence. It depicts a gridded pavement with a row of figures, all of equal size, marching into virtual distance, each figure stepping on a separate square (ill. 73). As the squares diminish in receding depth, each figure grows shorter in progression from the nearest in front to the furthest in back, although all their heads remain at level on the same horizon line. Interestingly but curiously, we see evidence that Pisanello projected his sight lines from an Albertian distance point across the central orthogonal, but then changed his mind, ignoring the guide marks on the central vertical, and drew the transversals in freehand.[6]

Finally, he indicated how the gridded floor allowed the artist to project the illusion of an arched architectural ceiling, simply by means of parallel

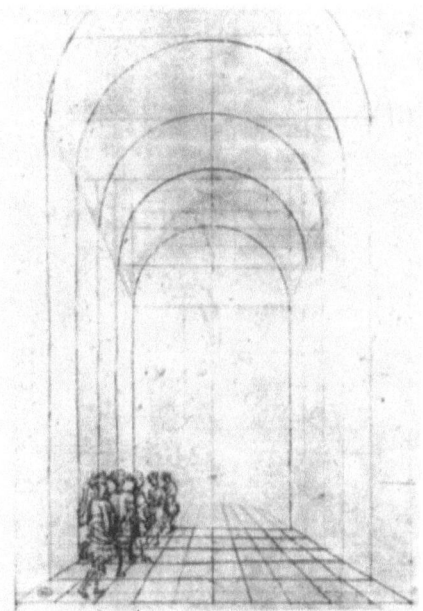

73. Antonio Pisanello, *Perspective Drawing*, ca. 1445. Musée du Louvre, Paris/Photoprint: Bridgeman Art Library.

semicircles, drawn upward from the ends of each receding transversal—suggesting, by the way, Masaccio's likely method for constructing the fictive vaulted space of his *Trinity* fresco.

Just how easy it must have been for artists to learn and employ Alberti's perspective method is evidenced by my own teaching experience. Every year in my "Art of the Italian Renaissance" class at Williams College, I have each of my students draw a picture of an interior room with door and windows, and people standing among chairs and tables set on the receding transverse squares of a gridded floor, all according to Alberti's rules. It takes approximately an hour for me to demonstrate with simple diagrams on the blackboard, and then have my students apply the rules to their own pictures. Few of my students (of art history) have had any studio training and none boast artistic talent, yet within the next two hours all are able to lay out the basic perspective structure of an Albertian *historia* just as competently as any Renaissance painter.

CHAPTER THIRTEEN

Alberti's Window

In both the vernacular and the Latin texts of Book 2 there is a description of a handy method for making a picture, Alberti's famous "window," conceived here as a rectilinear frame gridded with a network of strings. He called this *velum* in Latin, which refers to a fabric such as a loosely woven, transparent "veil." Below is a slight abridgment of Alberti's original words:

> One can not find anything more convenient than that veil, that I myself, among my friends, usually call the intersection, whose use I now discovered for the first time. It is of this kind: a veil woven of very thin thread and loose texture, dyed with any color, subdivided with thicker threads according to parallel positions, in as many squares as you like, and held stretched by a frame. Which [veil] indeed I place between the object to be represented and the eye, so that the visual pyramid penetrates through the thinness of the veil. The intersection of the veil in fact certainly offers no few opportunities, first of all because it always presents the same surfaces unchanged. After having established the limits, in fact, you will find on that place the original apex of the pyramid, a thing which is really very difficult without the intersection. And you have learned that in painting something how it is impossible to imitate correctly what does not maintain without interruption the same aspect of itself towards him who paints. . . . For when you observed over there in one parallel, the forehead, in the next the nose, in a nearer one the cheeks, and in a lower the chin and all things of that sort situated in their own places, in the same way there on the panel or a wall also subdivided by corresponding parallels you will have placed all the things in the best manner. This same veil finally is of great help in

perfecting a picture, since, in this plane of the veil, you perceive delineated and depicted an object prominent and rounded in relief.[1]

There is no illustration of this device in any of the known manuscripts of Alberti's treatise, but it was frequently depicted in later commentaries based on this textual description. Illustration 5 in chapter 1 is a sixteenth-century German woodcut of such an example. In any case, Alberti's latticed "window" remains his most original contribution to the art of painting. It also represents an attitude significantly different than Brunelleschi's. By this novel approach Alberti introduced, inadvertently, a secular alternative to perspective's original religious purpose.

Certainly, the notion of "window" indicates that Alberti was more concerned with confronting directly the physical "here" of the visual world (but not necessarily the "now" as will be explained shortly), and not its metaphysical reflection. In fact, his only reference to the mirror at all is that the painter should employ it as a technical tool for detecting drawing defects, still a helpful painters' practice to this day.[2] Furthermore, Alberti stated that the visual world beyond the window owed its beauty to a numinous *Natura*, as if referring to a personified pagan deiform rather than the Christian God (who is barely alluded to anywhere in his *Treatise on Painting*).

So why was the "veil" Alberti's most important contribution to the history of Western art? Although claiming he was the "first to discover" it, Alberti acknowledged that artists before his time customarily employed the grid as a means for blowing up smaller scale preliminary drawings to full size in their finished paintings—witness Masaccio's *Trinity*. What then was so unique about Alberti's gridded "window"?

First of all, it should be noted that the veil as such had nothing to do with perspective construction per se, nor was it simply a transfer device. The artist should employ it for copying his or her subject directly, literally mapping a portion of the visual world as if seen through a transparent cartographic chart and aligning the details according to a vertical (meridian) and horizontal (parallel) coordinate system. The artist was expected to know already how to compose his picture according to Alberti's perspective method, full instructions for which had already been explained in Book 1.

It is extremely important to understand this, because Alberti's window was intended as a further shortcut that eliminated much of the need for intricate construction drawing prior to setting up a picture, but only after the artist was fully aware of the author's elucidated optical principles. The geometric-coordinate system of the veil permitted the artist then to lay out the composition without having to measure every detail from the centric point or horizon line, assuming again, of course, that the artist had already thoroughly appropriated these optical laws into his or her second nature. In spite of Alberti's enthusiastic promotion, subsequent Renaissance painters took little advantage of it until a century later when Albrecht Dürer, in his widely circulated book *Underweysung der Messung* (Treatise on Measurement, 1525), showed in a famous series of woodcuts how the technique could be variously applied (ill. 74).

In effect, each square of the veil frames its own separate pictorial detail, all scaled exactly to one another and to the whole by virtue of its relative position within the coordinated grid. Once the artist fixes the veil in front of the subject to be painted, he or she can concentrate on painting the framed details one by one in any order, always confident that the perspective distortion of each object in the virtual space of the square will be automatically integrated with all the others. In any case, the veil did not really become an indispensable artists' tool until the seventeenth and eighteenth centuries, when exacting portraits and landscapes of private estates (suitably laundered of any unwanted blemishes) came more and more to be

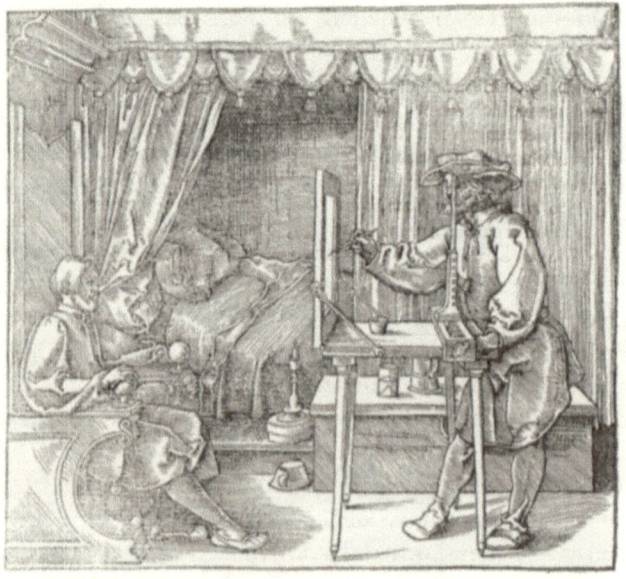

74. Albrecht Dürer, woodcut of an artist drawing with a perspective device based on Alberti's window, *Underweisung der Messung*, 1526. Chapin Library of Rare Books, Williams College, Williamstwn, Massachusetts.

commissioned by members of the nouveau riche mercantile class aspiring to acquire the patrician look.

Alberti's veiled window significantly differed from Brunelleschi's mirror in that it opened upon actual nature with no special sense of it being a mere reflection of God's divine paradise. Nonetheless, Alberti in his time was hardly interested in giving his artist readers a better means of copying what is seen in the purely "here and now" sense of secular reality. Rather, and even though he surely wanted *Natura* to be depicted the way we see her "here" on earth, he did not want the lady to be represented without her artificial makeup. Through the veil, the painter should not just behold "nature naturing" (*natura naturans*) in the wild, weeds-and-all sense, but "nature natured" (*natura naturata*); that is, manicured and modified according to the ancient understanding of moral order.

Alberti, like Antonino, referred to the subject matter of painting as *historia*. The word in both instances was intended to summon up didactic images of glorious ages past when men were nobler and their deeds more heroic. Indeed, it does seem likely that Alberti, like Fra Angelico, also began to realize that geometrical perspective didn't really work well if it only tried to evoke the ineffably divine. To be useful to painters at all, Alberti seems to have reasoned, shouldn't it better serve to promote secular human ethics rather than spiritual devotion—to impress and improve the present age by holding up before its eyes idealized models of those once and future virtues, especially as practiced in classical antiquity?

For Alberti, indeed, ancient Greece and Rome in timeless landscape represented the perfect setting in which to stage his moralizing *historia*. He especially hoped that his beloved Florence would become no less than Rome reborn. Florence's already famous artists, armed with the new perspective (which according to Filarete not even the ancients understood), should therefore set the didactic example. Indeed, artists, the presumed audience of Alberti's treatise, were addressed as if they were also humanists, as knowledgeable as Alberti himself concerning the great heroes and events of classical, and even pagan, antiquity. Alberti was truly the first Renaissance literary scholar to apply classical references to the visual arts, even before contemporaneous artists had any idea who "the famous daughter of Inachus" was, much less "Alexander's noble horse Bucephalus."[3]

The author of *De Pictura* cleverly let it be assumed that painters did recognize these deliberately dropped names. It was his way of encouraging them to learn about such erudite subjects by baiting their social-climbing ambitions. Like Antonino once more, Alberti not only desired that painting should be uplifting but that it should avoid being filled with distracting, extraneous

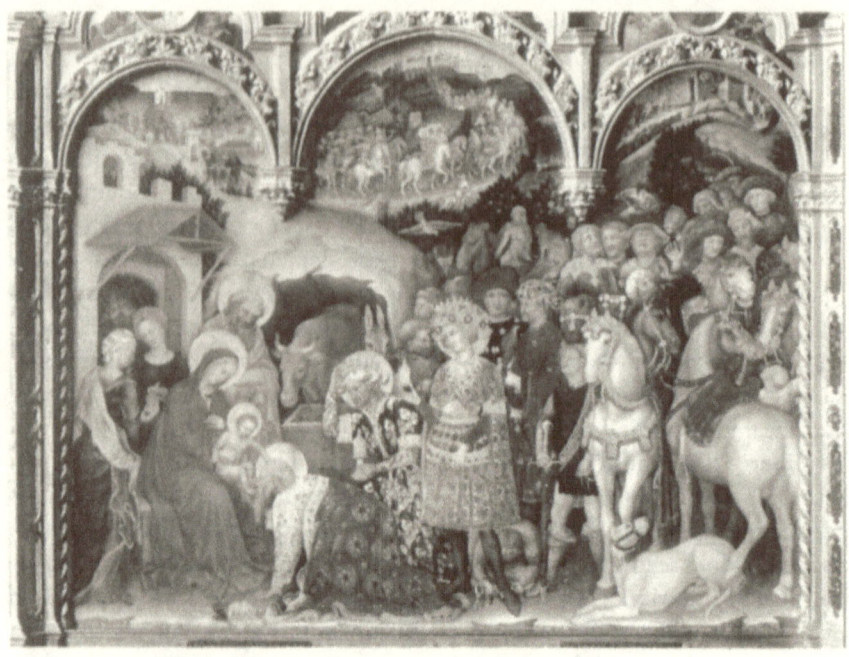

75. Gentile da Fabriano, *Adoration of the Magi*, Galleria degli Uffizi, Florence, 1424. Photo: Gabinetto Fotografico del Polo Museale Fiorentino.

details. He too would have disliked Gentile da Fabriano's *Adoration of the Magi* (ill. 75), especially for its overabundance of superfluous figures, which, as Alberti said,

> I blame . . . those painters, who in some way want to appear copious, or perhaps not to have left any space empty; for this [reason] they do not follow any composition, but scatter everything in a confused and illogical way; from this the *historia* does not seem to follow an action but generates confusion. . . . As in fact a few number of words produces majesty in a prince . . . so an appropriate number of bodies in a *historia* confers dignity.[4]

Moreover, in his own time Alberti's views on painting would not have been considered counter even to those of Antonino. The rediscovered literature of ancient Greece and Rome was becoming de rigueur reading for every intellectual in late medieval Italy, including members of the Church hierarchy. Even though many of Alberti's references were to pagan mythology, Christian theologians as well as humanists recognized in these antique stories the same virtuous qualities of a lost but more decorous Golden Age, nearer in time to God's Genesis. Possibly in that ancient time some of God's original intentions were still preserved although obscured under heathen veneer, nevertheless just waiting to be exposed again through proper scholarship and interpretation.

Furthermore, some of Alberti's classical references were thinly disguised attempts to upgrade traditional Christian subjects by recasting them in the form of antique images. For instance, when Alberti described how to paint the dead Meleager, hero of the Trojan War, being borne away by "those who . . . appear to be distressed and to strain with every limb," he was inferentially advising painters on how to improve their renditions of the dead Jesus being taken down from the cross. Alberti may actually have been thinking of Fra Angelico's *Deposition* (ill. 65), which had been recently put up alongside Gentile's *Adoration* in the Florentine Church of Santa Trinita. As noted in a previous chapter, he was no doubt familiar with that masterpiece and would have approved of its counterbalancing *dignitas*.

In fact, Alberti's verbal description of the fatally wounded Greek hero, along with Fra Angelico's painted image of the dead Christ, quickly furnished the canonical representation of the Savior's body, limp like the fallen Meleager, with his muscular, pronated arm "drooping inertly down" over the edge of his burial shroud. Nevertheless, Alberti really intended that his perspective should enhance the moral rather than the mystical message of the Christian stories, and all the more eloquently if evoked from the hallowed *historiae* of ancient Greece and Rome.

Alberti therefore urged the painter to begin painting such an antique event by first laying out his abbreviated foreshortened *pavimentum*, and on this optical lattice outline the projected composition. Next the painter should set up the veil and copy a natural landscape, fitting the desired details to the gridded squares to serve as framing background to the scene already in perspective. Into this "window" setting tangible human figures should be inserted, each in a varied but dignified pose signifying a virtuous gesture in the heroic story. Alberti clearly believed that such noble subject matter appropriately represented through idealized human interaction even in a modern landscape provided just as much ethical guidance as did the traditional biblical allegories. As Anthony Grafton noted, "Alberti, in short, committed himself throughout his life to making classical texts, ideas, and forms live in what he recognized quite happily as a non-classical world."[5]

On the other hand, implicit in the linear structure of Alberti's optical model was also the *Urform* of the universe as conceived in the Creator's mind at Genesis, which was then projected into the void. Even if Alberti may have felt that perspective geometry did not manifest God's postlapsarian mysteries very convincingly, he did intend that artists who employed his perspective method could now at least create a replica of Nature's primeval world framework. The geometrically regulated, virtual space behind Alberti's window should oosmetically imbue images painted in it with the same intrinsic harmony and order that God intended mankind to live by on earth before the Fall, which, in the metaphoric form of *historia*, should be communicated back to the viewer standing before the picture.

Finally, if Alberti eschewed Brunelleschi's mirror, he did borrow the other unique idea of his perspective predecessor, namely the little peephole to concentrate and thus dramatize the viewer's single eye-point, but this time not through the back of the picture but in front of and aimed directly at the picture itself. Actually, the hole went through one end of a closed perspective box, or *camera ottica* as it came to be called, with the other end open and held against a light source. Here, a picture painted on a translucent surface should be inserted, which resulted, as the inventor enthusiastically proclaimed, in his viewers experiencing a "miracle of painting." Although it was not mentioned in *De Pictura*, Alberti's *camera ottica* was described in the so-called *Vita anonima* (Anonymous Life), which perhaps was an autobiography written by Alberti in the mid-fifteenth century:

> By looking into a box through a little hole one might see great plains and an immense expanse of sea spread out till the eye lost itself in the distance. Learned and unlearned agreed that these images were not like natural things but like nature itself. These demonstrations, as he [Alberti] called them, took place by night and day. In the former, you saw Orion, Arcturus, the Pleiades, and other shining stars, and the moon rising above high mountains; by day you saw the blaze of dawn as Homer describes it. Certain Greeks, famous men and skilled seafarers were astonished when he showed them, in his little world, a ship far out to sea. "Now it labors in the tempest," he said, "But tomorrow you will find it in the harbor."[6]

Do we have here an early intimation of the optical telescope, actually a "perspective tube" just like Alberti's *camera ottica*, the "window" effect of which is simply extended and expanded by magnifying lenses?

CHAPTER FOURTEEN

Alberti's Legacy

Raphael's Stanza della Segnatura *and Beyond*

Pope Julius II della Rovere (1443–1513), elected to the Roman Church's highest office in 1503, so abominated his predecessor (by two), Alexander VI Borgia, that he refused to inhabit the regular papal quarters in the Vatican Palace, even though it had been recently redecorated at great expense by Bernardino Pintoricchio (ca. 1454–1513). Instead he decided to move to another suite on the floor above, ordering it covered with new scenes more appropriate to his own taste and ambitions. Here in 1508 was summoned the twenty-five–year-old painter Raphael Sanzio of Urbino (1483–1520) to create in fresco what would quickly be regarded as one of the greatest set of masterpieces in the history of Western art. Raphael's style, especially as evinced in this grand commission (in which several assistants participated), represented, more than that of any other independent artist or workshop, the ultimately accepted, most popular approximation of what Alberti might have imagined as idealized *historia* seen through his "window."

Although Raphael died in 1520, at the tender age of thirty-seven, a full two decades before Alberti's *De Pictura* was finally printed (that is, before Alberti's unillustrated ideas began to reach a worldwide audience), the young painter had already absorbed the full essence of Alberti's notion of *dignitas.* Thus it might be said that, ever after, anyone reading *De Pictura* would regard the paintings by Raphael, especially the murals of Pope Julius's freshly frescoed first apartment, the *Stanza della Segnatura* (Room of the Curial Tribunal), but actually intended to be the repository of the Pope's private library[1]), as synonymous with Alberti's ideas (ill. 76).

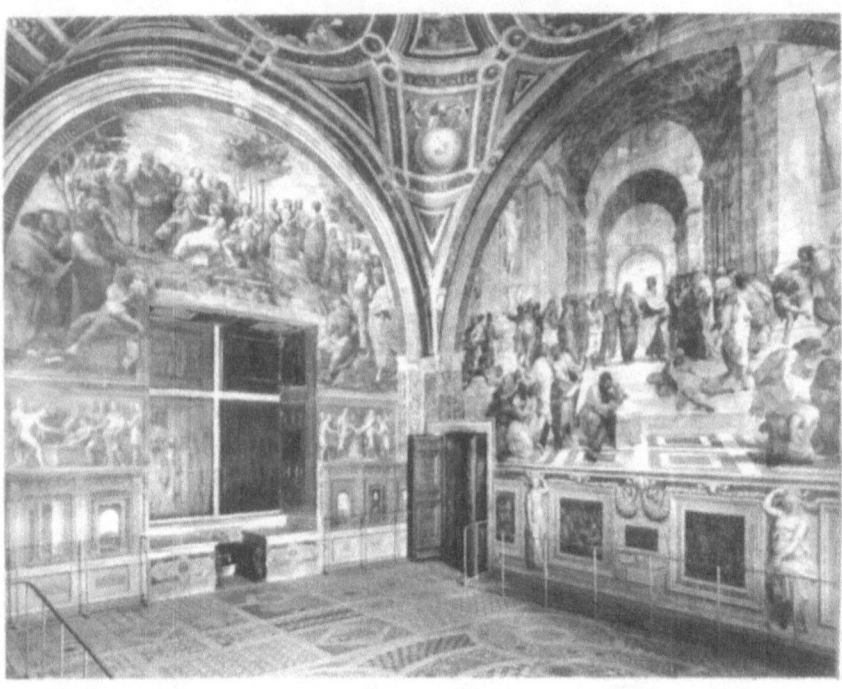

76. The *Stanza della Segnatura*, Palazzo Vaticano, Vatican City.
Photo: Alinari/Art Resource, N.Y.

The subjects chosen for the frescoes on the four separate walls were to be metaphorical depictions of the standard bibliographical categories in the late Middle Ages: *Theology, Philosophy, Jurisprudence,* and *Music and Poetry.* Raphael represented them in the most carefully studied style of revived classical antiquity, yet so effortlessly and unself-consciously that Alberti would surely have proclaimed them "miracles of painting."

The city of Rome, the ancient capital of that idealized civilization whose revival Alberti so fancied in Florence, had quite fallen into decay since the times of the Caesars. During the Middle Ages it was a veritable slum, its once glorious architecture in ruins, inhabited by transient squatters by day and wolves and muggers by night, and many of its once elegant sculpted monuments deliberately smashed and used for ignominious mortar fill.

Only with the return (from Avignon) of the Papacy in the fifteenth century did the city again recover its noble heritage, especially as the popes began to see themselves as latter-day Roman rulers, even adopting the ancient pagan high priest's title, *pontifex maximus.* By the time of Raphael's arrival, antiquarianism in Rome was in high fashion, already encouraged by the earlier writings of the humanists, but now joined by artisans eager to explore the ruins and discover long-buried classical artistic motifs. The more they found, the more they sought to learn the long-forgotten techniques and advertise themselves as antiquarian experts. Rich patrons in the entourage of the newly empowered popes just as eagerly sought the services of these skilled artisans to have their own dwellings decorated like antique Roman palaces. More than ever, Italian artisans really needed to know something about "Alexander's horse, Bucephalus" just as Alberti had earlier prompted. In 1515, Raphael himself was appointed by the Pope as director of all the antiquities excavations being carried on in and around Rome.

   Especially popular among the revived classical techniques was the art of imitating low sculptural relief, the way ancient carvers and plaster workers depended on light and shadow to provide the illusion of perspective, but not of deep space. Rather, the old artists seemed to have been more interested in creating the illusion of perspective protrusion, as if the figures in the sculpted scene were stepping forward into the viewer's space, for example the sculpted scenes on the first-century AD Arch of Titus in the Roman Forum (ill. 77). Not only the sense of protrusion but the procession of figures across the foreground plane rather than receding behind it, again as in this antique relief,

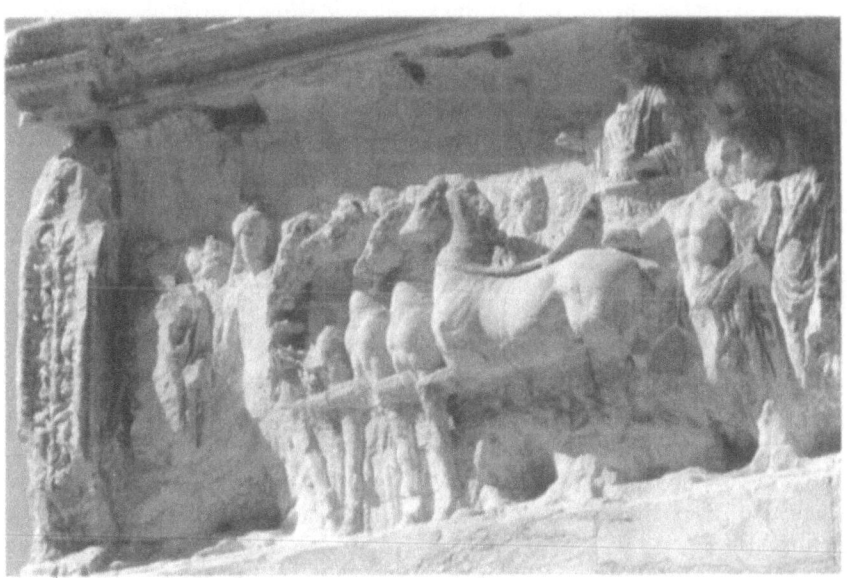

77. Stone relief, *Spoils of Jerusalem*, Arch of Titus, Rome, 81 AD. Photo: author's collection.

increasingly attracted sixteenth-century Roman artists to apply these same compositional arrangements.

Alberti had already admonished painters to achieve the effect of relief in painting by favoring black and white over too much display of other colors, as he wrote in *De Pictura*:

> Without doubt I affirm that the wealth and variety of colors are greatly useful to the grace and beauty of a picture. But, this is what I would want: let the prepared painters consider that the highest quality and mastery reside only in the distribution of black and white and that, in having to place accurately these two, one must devote all talent and zeal. As, in effect, the incidence of lights and shadows comes into sight in which place surfaces swell up, or where they shrink by hollowing out, or how much every part moves aside or strays, so the distribution of black and white produces what became praised in the Athenian painter Nicias, or what an artist must greatly look for: that his painted objects appear very much to protrude. . . . Certainly I will consider insignificant or mediocre that painter who does not understand clearly how much power every shadow and light produces on each surface. With the consent of both the ignorant and the expert, I will praise those portraits that seem to protrude as sculpted from pictures.[2]

Indeed, the science as well as the art of rendering chiaroscuro became of much concern in the sixteenth century, particularly as the advancing technology of printmaking made it possible to reproduce the most subtle modeling effects in book illustrations. More and more technical manuals began to be published on the various applications of shading and shadow casting, for instance in the representation of multifaceted geometrical solids, as in the cutting of gemstones.[3] Illustration 78 is such an example from the Nuremberg jeweler Wenzel Jamnitzer's 1569 *Perspectiva corporum regularium* (Perspective of Regular Bodies).

Furthermore, as Marcia Hall has shown, a conscious "relief-like style" of painting was much practiced by Raphael's followers after his untimely death in 1520.[4] As mentioned, this style also appealed to the papal aristocracy who regarded themselves as patrons of the revived Roman *Imperium*. This was a time of much concern in Catholic Rome as the Church increasingly felt the need to stress its ancient Roman heritage, and thus its divinely decreed universality, to help quell the already early stirrings of religious reformation.

Returning now to Raphael, let us examine his most famous masterpieces in the *Stanza della Segnatura* on the two broad side walls flanking the true

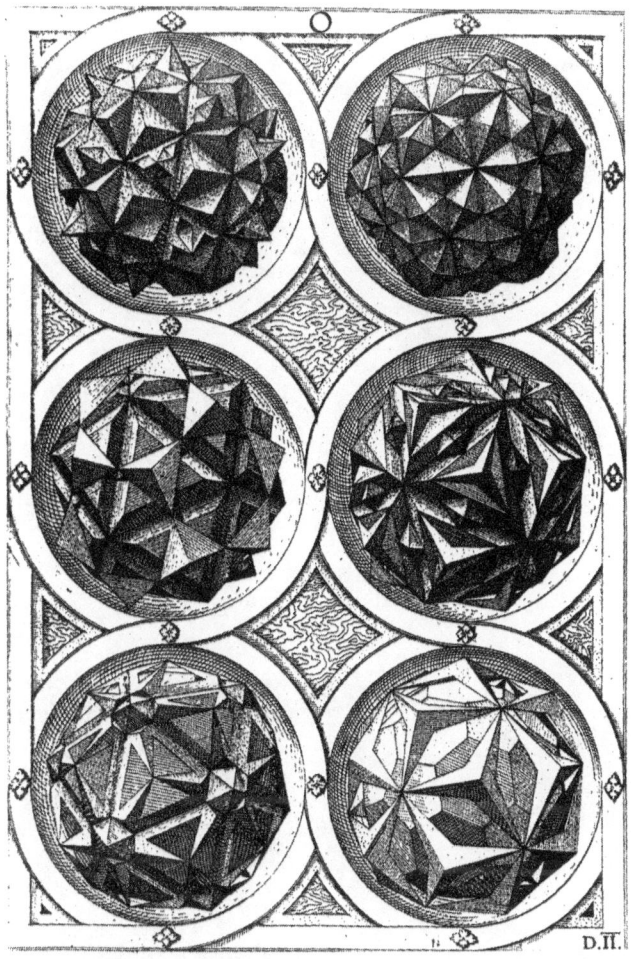

78. Wenzel Jamnitzer, engraving of cut jewels, *Perspectiva corporum regularium*, Nuremberg, 1568.

windows, the so-called *Dispute of the Sacraments* and *School of Athens* opposite, actually the modern names for the original literary themes, *Theology* (ill. 79) and *Philosophy* (ill. 80).

Raphael began these frescoes by composing each upon a carefully constructed Albertian *pavimentum* converging on a centric point. Moreover, Raphael took advantage of the centric points to emphasize the respective symbolic themes of each. In *Theology* the point is coincident with the base of

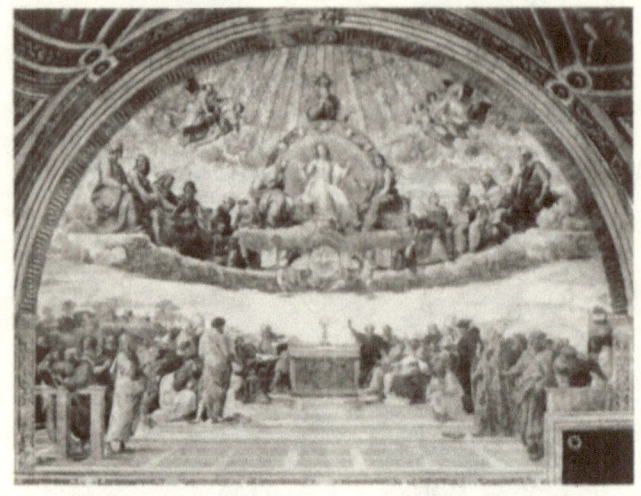

79. Raphael, *Theology, Stanza della Segnatura*, Palazzo Vaticano, Vatican City, 1509. Photo: Alinari/Art Resource, N.Y.

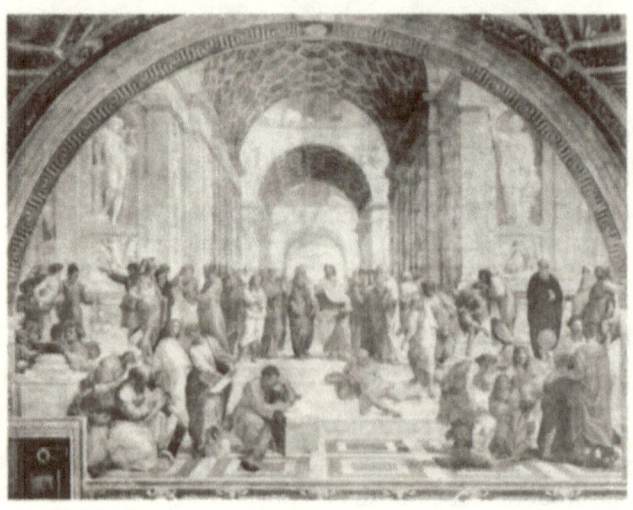

80. Raphael, *Philosophy, Stanza della Segnatura*, Palazzo Vaticano, 1510. Photo: Alinari/Art Resource, N.Y.

a circular monstrance containing the sacred Host, and in *Philosophy*, just as appropriately, the point falls between Plato and Aristotle standing under the painted arches, exactly at the level where these two most illustrious thinkers of classical antiquity elegantly hold their important books. The unaffected decorum that Raphael was able to express in the idealized stances of these engaging figures came to be especially admired in Italian courtly circles. Called *sprezzatura* in Italian, it became the required manner in which people of gentility should actually comport themselves in public, perhaps best translated in English by our modern vernacular street term, "by looking cool."

Plato holds his *Timaeus* in which he describes God's primal vision of the universe and with his other hand points to the sky. Beside him, Aristotle holds his book, the *Ethics*, which concerns human moral behavior. With his other hand he points to the earth. Indeed, the philosophers are here discussing the lessons of their two great treatises. One can almost hear Plato reaffirm the meaning of truth, just as Fra Antonino and even Alberti would have understood it:

And knowledge is of two kinds, one turning its eyes towards transitory things, the other towards things which neither come into being nor pass away, but are

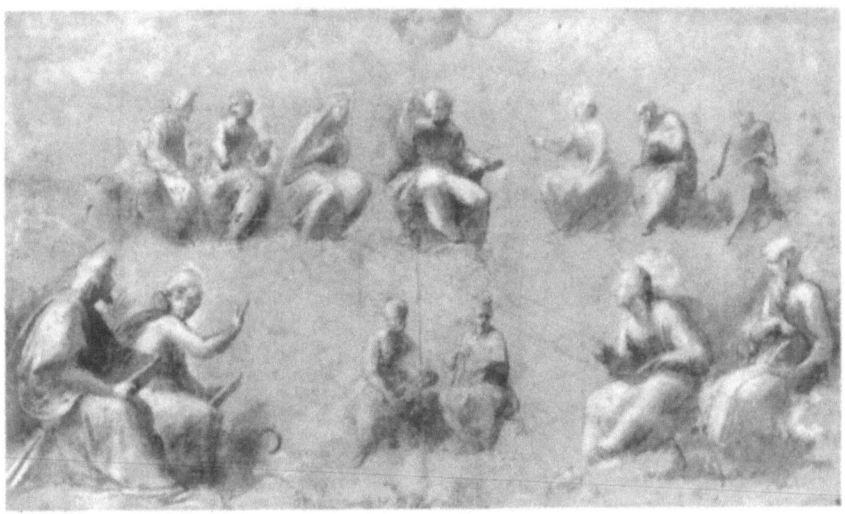

81. Raphael, chiaroscuro sketch for *Theology*, 1509. Ashmolean Museum, Oxford.

the same and immutable forever. Considering them with a view to truth, we judge that the latter is truer than the former.[5]

Raphael deliberately interrupted the Albertian recession in both of these frescoes with raised platforms running across the pictorial middle ground on which rows of figures are arrayed more or less parallel to the picture plane. They are not yet quite in the projected "relief-like" style, but Raphael has certainly paid attention to Alberti's admonitions about the need to emphasize their sculptural quality by modeling them carefully in chiaroscuro before applying color. This is very evident in the many black and white drawings he made in preparation for these paintings. Here is one he composed, and then rejected, for *Theology*, showing two parallel rows of figures modeled as if illuminated by a raking light and projecting forward like carved relief from a flat background (ill. 81).

Clearly Raphael was toying with a "relief-like" effect in this early composition. In fact, *Theology* was the first of the four frescoes to be painted, and in this drawing he hadn't yet realized his eventual brilliant idea to take advan-

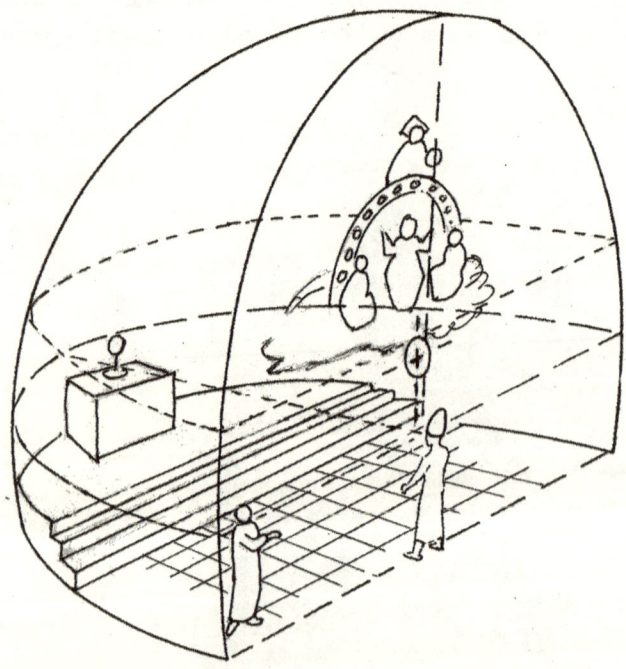

82. Spatial reconstruction by author of Raphael's *Theology*.

tage of the curving arch around the wall, and imagine it as a "window" into a quadrant of a sphere.[6]

In the completed fresco, Raphael chose to depict the same two groups of figures, Apostles raised up on a heavenly cloud and famous historical personages on earth below seated and standing side by side in a row seeming to parallel the one above. Each row is a curve, the latter viewed as if seen from earthly eye level, while the upper is viewed from below. The whole composition has the effect of a diorama, with three-dimensional figures on a semicircular stage set against a painted landscape backdrop.

But notice how cleverly Raphael applied Alberti's window vision of an idealized geometrized classical world in order to reaffirm the reigning Pope's reinvigorated confidence that he was the sole interpreter on earth of God's word in heaven. First of all, Raphael positioned the altar supporting the monstrance containing the Host in a deliberately ambiguous alignment between the upper and lower rows of figures. The spatial disparity becomes even more confusing if we compare the sizes of Jesus, Mary, John, and God the Trinitarian Father with the Apostles seated immediately in back, and also with the figures standing below on either side of the altar. Although they are much larger than the Apostles in back, they are exactly the same size as the foreground figures at both ends of the two curving rows. In other words, unless Raphael uncharacteristically intended these to be iconically larger than normal living humans, he has actually placed them far forward spatially. Furthermore, the cloud that supports the Trinitarian grouping should be understood not as another parallel curve but as a line segment of a chord actually forward from and stretching straight across the great transverse arc behind and almost tangent with the picture plane itself. Also nearly flush with the picture plane is the encircled Holy Ghost on a vertical axis below Jesus's throne. The artist thus positioned the Holy Trinity not just above the altar but quite before it, hovering over the empty pavement between the gesturing interlocutors on either side in the near foreground, as in illustration 82, my reconstruction.

Whatever the reason, our ingenious painter made it possible for his patron, Pope Julius II, to imagine himself invited into the fictive space of the picture, beckoned by the young man depicted at the left.[7] How could the Pope not resist the urge of Raphael's extraordinary illusionism, to step forth and greet his late kinsman, the towering beardless figure wearing a tiara depicted to the right, that has always been considered a portrait of his uncle and earlier predecessor, Sixtus IV (1414–84). Pope Julius could even envision himself advancing toward the painted altar, his own, since Raphael inscribed the Pope's name on it not once but twice.[8] There, a few steps before the sacred monstrance, the Pope could feel himself connected meta-physically to the divine axis of the Holy Trinity, confirmed by Alberti's perspective as crucial link between triune God in heaven and mortal man-kind on earth.

Questions are provoked, however. Did Raphael's art, in effect the ulti-mate expression of Alberti's *historia* as well as his perspective window here reapplied to a profound theological subject, really help in sustaining the sa-cred doctrine of the Roman Church? Or did his geometrized, classicized vision of a holy hierarchy posed like corporeal Olympian deities uninten-tionally undermine it? What of Raphael's Euclidian representation of the spherical cosmos showing both ineffable heaven and effable earth quite joined together in one uniform spatial continuum? Even his brief representation of a golden empyrean over the head of God is framed by what appear to be ra-diating meridians indicating that the divine realm itself is geometrically contained. Might not such a material concept as depicted here, in which all space is subject to the same mundane perspective mathematics, fly in the face of the Church's own insistence that earth and heaven exist in separate media, and that the former is but a pale reflection of the latter? Would not Fra An-tonino have indeed complained that light streaming in from Raphael's gilded empyrean at the top of the fresco should at least be shown as refracted when it enters the lowly medium of earthly atmosphere? Finally, how much effect does any pictorial representation at odds with the Bible especially when so aesthetically attractive, have on popular opinion, particularly if approved by the Pope for his own private library?

In spite of the pictorial contradictions, not only the sixteenth-century public but many contemporaneous artists quite admired Raphael's daring composition. Titian's great *Assumption of the Virgin*, painted for the Church of the Frari in Venice in 1516/18, is perhaps the most splendid example of Ra-phael's spreading influence (ill. 83).

On the other hand, the real indicator of just how impressively such a po-tent image can translate into popular acceptance is not so much revealed in the art of great masters as it is in how quickly it becomes a convention of minor talents, the run-of-the-mill artists hired by less discriminating patrons who desire to have their commissions picture the latest fashions but usually in rather insipid interpretations.

A telling example of what Raphael's influence wrought at this banal level is a sixteenth-century version of the ever recurring "ladder-to-heaven" ap-parition, *The Vision of the Blessed Amedeo Menez de Sylva*, by the Spanish painter Pedro Fernandez (known in Italian as Pietro Ispano) who lived in Rome during the years around 1514, actually working for Raphael in the

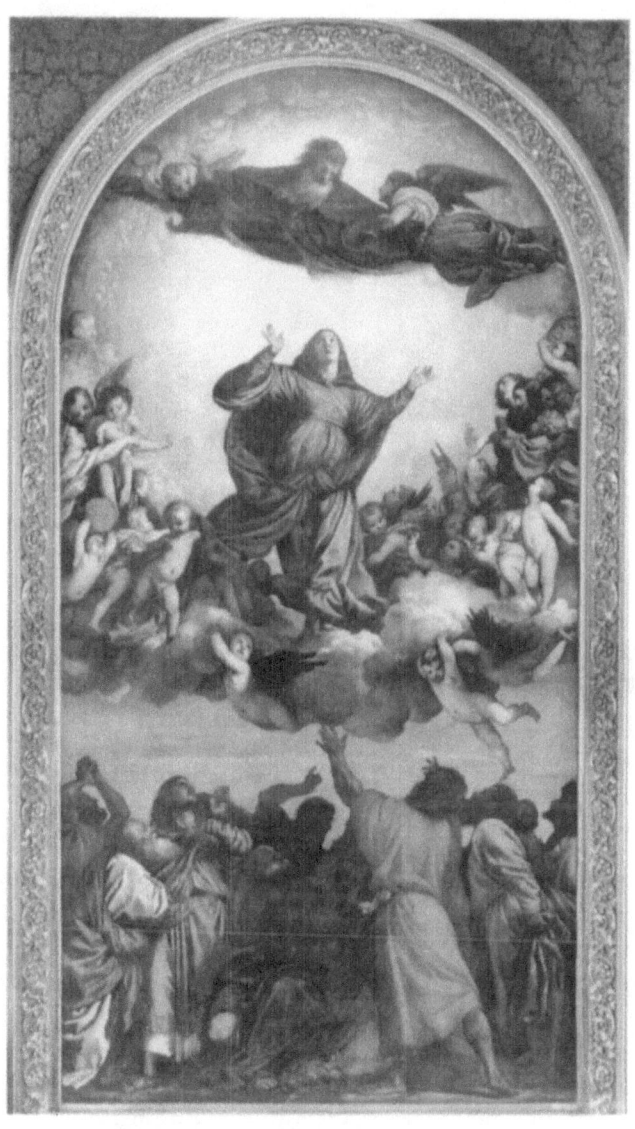

83. Titian, *Assunta*, Church of Santa Maria degli Frari, Venice, 1516. Photo: Alinari/Art Resource, N.Y.

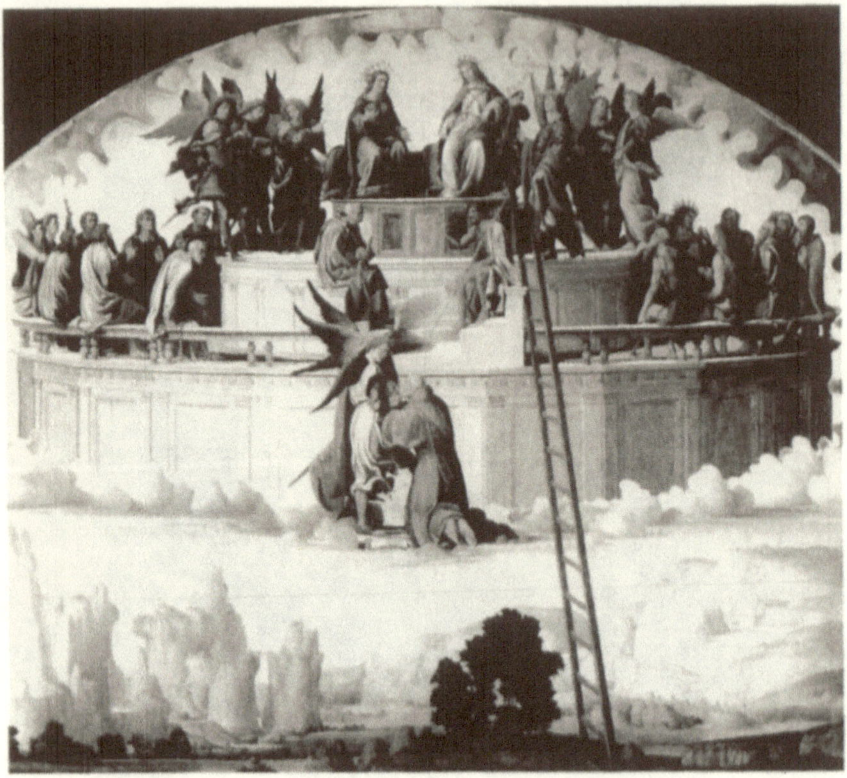

84. Pedro (Pietro Ispano) Fernandez, *The Vision of the Blessed Amedeo Menez de Sylva*, Galleria Nazionale d'Arte Antica, Rome, 1514. Photo: author's collection.

decoration of the Vatican apartments. Pedro's large panel (9x10.5 feet), so derivative of Raphael's *Theology*, was originally commissioned for a monastery church outside Rome (ill. 84). It depicts a fifteenth-century Franciscan mystic and semisaint who claimed to have had a miraculous dream in which he visited the celestial court in heaven, including a personal audience with Saint John the Baptist.

As is embarrassingly obvious in Pedro Fernandez's picture, the artist has imagined the holy man's heavenly visit as taking place on a three-tiered stage (in the currently popular Bramantine architectural style) perched on cardboardlike clouds as if suspended above an airy local landscape. Beato Amedeo supposedly arrives by arduously climbing up from earth on a long ladder much too rickety looking to have inspired any spiritual and/or moral transcendence. Viewers of this utterly objective imagery could only feel relief that the saint got safely to the top without the ladder collapsing.

This brings us to a brief review of just how dramatically the pictorial arts had changed after hardly a hundred years of increasing subjection to the rules of *perspectiva*. Not only the arts but even popular perception of the very Christian dogmas the arts were supposed to portray had changed.

Compare, for instance, Pedro Fernandez's uninspiring although geometrically logical concoction to another similar example of a "ladder-to-heaven" painting still very much in the preperspective tradition of the early Middle Ages, *The Vision of Saint Romuald*, by a provincial Italian master known only as "Pseudo Jacopino," dated circa 1375 (ill. 85).

The picture shows the eleventh-century saint asleep on a step before an outdoor altar dreaming of white-robed monks ascending to heaven on a ladder. Actually it was not the saint but a certain Count Maldolus who had the vision as he was sleeping in one of his fields. So moved was he that he deeded that inspiring site to Saint Romuald who then founded a religious order there, that of the famous Camaldolites. At the foot of a mountain nearby, Romuald erected his first monastery, here succinctly signified by the rose-windowed façade to the right beside the terraced hill on the left. The visionary ladder is set against a flat golden background, but leaning upon the edge of a blue-colored, star-speckled separated space at the top of the panel is the empyrean portal to which the climbers ascend. Not only do these simple signifiers articulate the ethereal message of this quaint story clearly, but, by their very abstractness, transport the viewer from the physicality of this world to the metaphysical ineffableness of the next. No matter how "unrealistic" and artistically naïve the traditional symbolism of this image, it still conveys a spiritual message far more sincere and emotionally moving than the improbable "vision" of Pedro Fernandez.

To be sure, after the assumption of linear perspective such mystical narratives had to be most unsubtly recast, their messages no longer conveyable by abstract symbols but rather by tangible people doing rational things in natural settings just as if one viewed them in real life through an open window (Alberti particularly eschewed the use of gilding to color background skies). Hence the value of the gridded veil: the artist need only position it before a contemporaneous scene, suitably manicure the natural landscape behind, dress the figures in bedsheet togas, arrange them in decorous poses, and compose all together according to correct perspective. In other words, Alberti's formulaic *historia* managed in the long run even to reduce holy mysteries to looking more and more like secular events. If Brunelleschi's mirror first introduced *perspectiva artificialis* in order to demonstrate how the

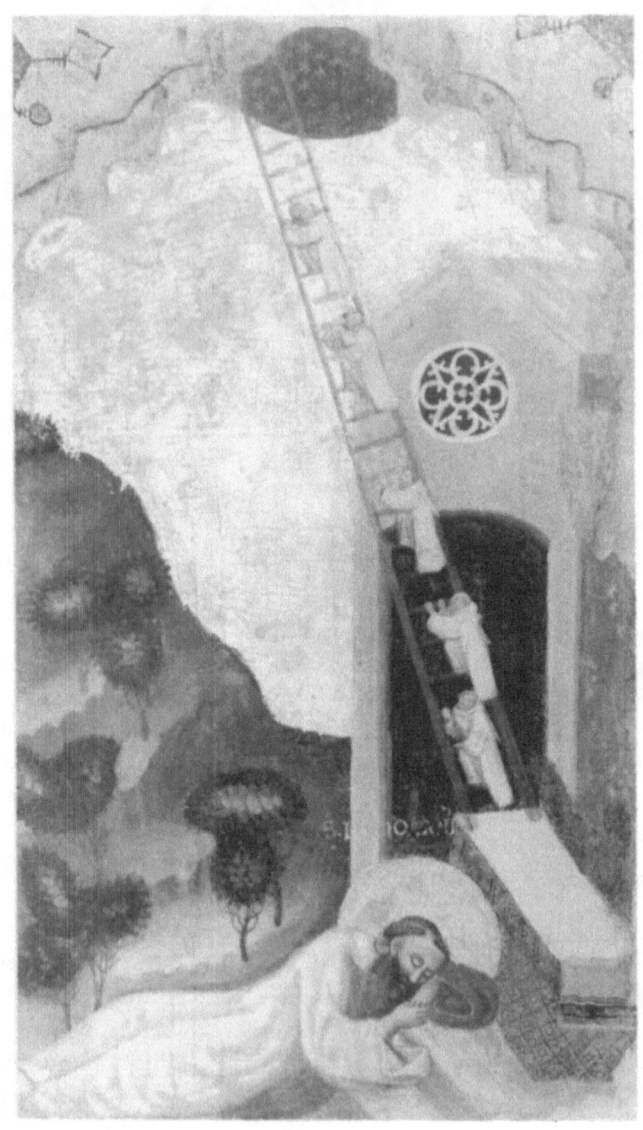

85. Pseudo Jacopino, *The Vision of Saint Romuald*, ca. 1375. Pinacoteca Nazionale di Bologna.

mundane world enigmatically reflected heaven's image, Alberti's window, even as ingeniously applied by Raphael, certainly, if only inadvertently for the next several centuries, compelled most European artists whenever they attempted to depict heaven to show it in the same image as the mundane world.

Interestingly, even as Alberti's influence was incubating in much of Western Europe during the sixteenth century, there also appeared a strong counterreaction in the form of a popular, almost antigeometrical style known in Italy as *la maniera*, "Mannerism," particularly inspired by the dramatic art of Michelangelo Buonarroti (1475–1564). Unlike Raphael's easily adaptable style, however, Michelangelo's riveting *terribilità*, while frequently imitated, never evoked the same sublime effect as by the master. Nowhere was this Mannerist reaction, especially with its hints of revived Gothic abstraction, more pronounced than in Counter-Reformation Spain, where many painters not only shied away from perspective but even abjured the use of bright colors. No edition of Alberti's treatise was published in Spain until the eighteenth century, and even in the current Spanish translations of Sebastiano Serlio's widely influential *Treatise on Architecture* for instance, Book 2 on the latest geometrical applications of linear perspective was deliberately left out.[9]

Furthermore, in the Spanish colonial Americas, especially Mexico, where Roman Catholic friar missionaries had founded art schools to teach native Indian artisans the rudiments of Renaissance art so that they might create appropriately spiritual Christian pictures as decorations for the new churches and *conventos*, perspective geometry was again clearly minimized. One of the more ardent promoters of how properly to instruct the Indians in this respect was a Franciscan friar named Diego Valadés who published a book in Latin on the subject of proselytization entitled *Rhetorica Christiana* (Perugia, 1579), which included a number of engravings by his own hand. Two of these, illustrations 86 and 87, are remarkably revealing of the traditional sentiment among the sixteenth-century Observant clergy that hearkens back, both to the influence of Fra Antonino (by now canonized as Saint Antonine; a printed copy of his *Summa theologica* was brought to Mexico in 1541) along with the never quite forgotten worry that too much emphasis on exacting geometrical perspective in religious images might undercut their spiritual message.

The first of these two prints by Valadés is known as the *Pagan Philosopher* (ill. 86), and shows a tall bearded figure standing on a gridded pavement that more or less recedes according to the Albertian rule. However, he is utterly preoccupied with geometric tools (3), spending his entire time measuring the

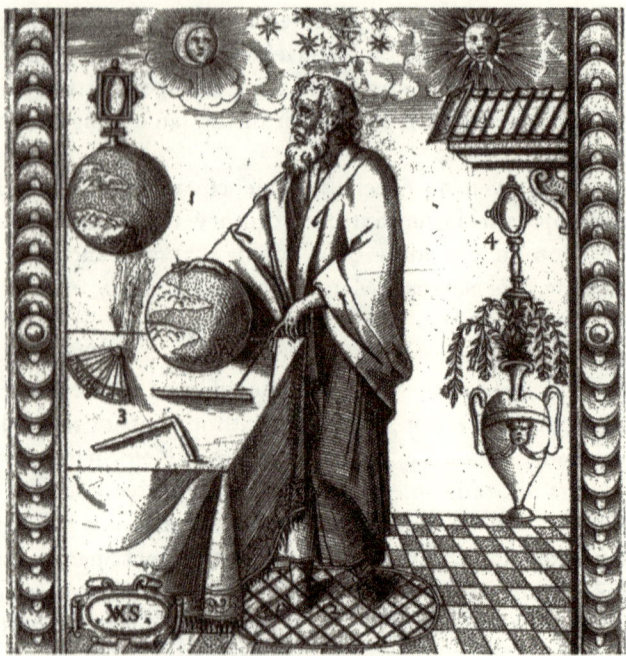

86. Fray Diego Valadés, *The Pagan Philosopher*, from his *Rhetorica Christiana*, Perugia, 1579.

earth (1). According to Valadés he thinks only of material things and sees in mirrors (4) only reflections of the transient present, which this time prevents him from foreseeing his eternal future in the hereafter. Examining the print even more closely, we notice that the pagan philosopher's feet are quite entrapped in another grid marked 2, a "veiled" warning that too much emphasis on perspective is dangerous to the would-be artisan in the service of the Church.

Valadés's antidote to the former's "paganism" is his companion engraving called the *Christian Philosopher* (ill. 87). Here the savant (A) is seated with his hand raised to his brow in heavy concentration. The corporeal earth (C) is beneath the desk, out of sight and under his feet as it should be, while a winged figure, his guardian angel (B), urges him to contemplate in his mind's eye only spiritual images converted "into figures" (E) just as Saint Antonine advocated. At the far right of the engraving at D is a mirror image of the Crucifixion that stands atop a skull. Valadés emphasizes that this

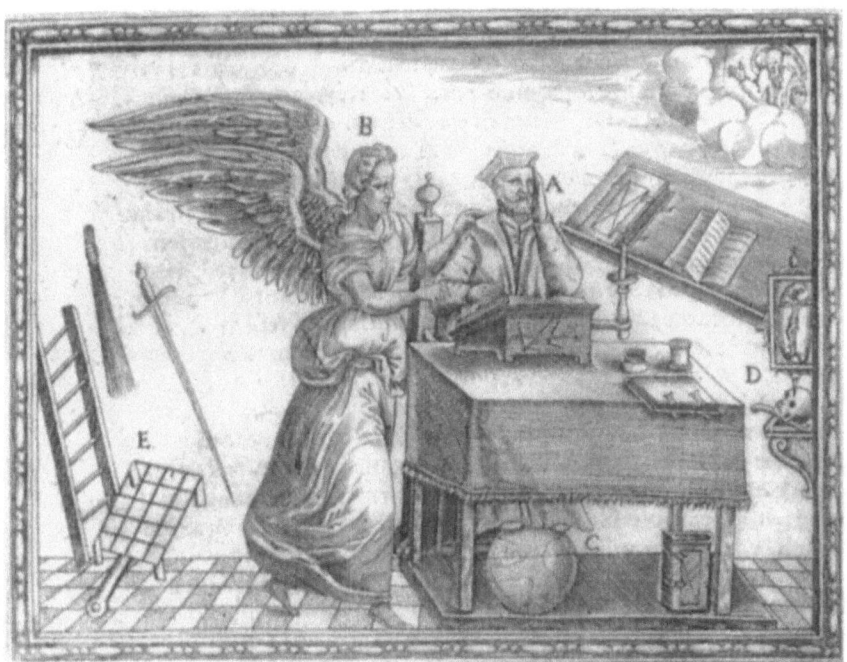

87. Fray Diego Valadés, *The Christian Philosopher*, from his *Rhetorica Christiana*, Perugia, 1579.

alone is the "mirror of nature," reflecting only truth and obscuring nothing.[10]

Even as the friars eschewed linear perspective, however, they were aware of the au courant "relief-like" painting in Italy, if not its "dark manner" counterpart at the same time practiced in Spain, and how these popular European styles might enhance their use of pictures as teaching tools for converting the Indians. In fact, most of the surviving murals painted by their Indian trainees in the various religious structures the friars had built for Christian religious purposes are colored mainly in black and white.[11] Their "relief-like" effect was intended to astonish native beholders into believing the depicted Christian events were happening in their very own space, "divinely present" as it were, and thus more "real" even in the spiritual sense than the traditional, flat, and more abstracted imagery of indigenous religious art.

A fairly typical example is a mural painted about 1570 probably by a converted Aztec Indian on the wall of the upper corridor of the Augustinian

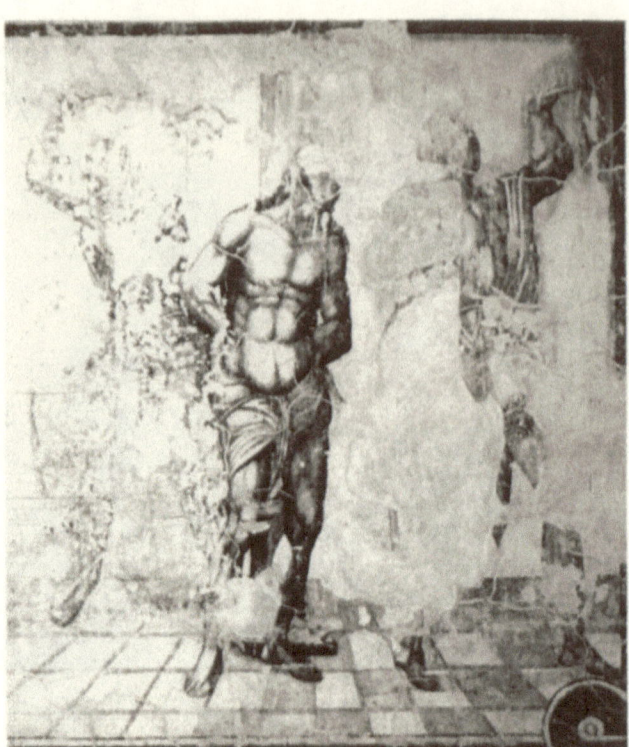

88. Anonymous Native American artist, *Flagellation of Christ*, mural in the Convento San Agustin, Acolman, Mexico, ca. 1570. Photo: Jorge Pérez de Lara.

convento of San Agustín Acolman, near Mexico City, depicting the *Flagellation of Christ* (ill. 88). Whatever the aesthetic quality of this unfortunately much abraded work, one can only admire the indigenous painter's newly trained ability to render chiaroscuro almost as competently as any Renaissance European artist.

CHAPTER FIFTEEN

Galileo's "Perspective Tube"

On the other hand, whatever negative effect Alberti's perspective may had on the "spirituality" of Christian art, and whatever didactic benefit it may have had on the morality of Western secular culture (the jury is still out on this), its strictly geometrical reasoning about the uniform nature of space did have some important relevance to the extraordinary astronomical revelations of the sixteenth and seventeenth centuries.[1] In circumstantial fact, just as Alberti's *De Pictura* was being printed in 1540, the Polish astronomer Copernicus (1473–1543), who had studied in Italy, was likewise thinking about uniform space organized geometrically. Based on his calculations of the concentric circles representing the orbits of the planets, he realized that the sun, not the earth, should be the center of our system (described in his treatise *De revolutionibus orbium celestium*, "On the Revolutions of the Heavenly Spheres," published in the year of his death).

In 1577, the Danish astronomer Tycho Brahe (1546–1601) observed a comet of that year traveled through the skies without shattering the so-called crystalline spheres traditionally believed to separate the celestial zones. Thus he affirmed that there were no impenetrable spheres, and that all the heavens above the moon were joined in an open continuum. Incidentally, Tycho was himself an artist and even invited Italian painters to decorate his private observatory.

Finally, and not surprisingly, the birth of Galileo Galilei in Tuscan Pisa on February 15, 1564, just three days before the death of the great Michelangelo in Rome, has given rise ever since to speculation that there must have been some kind of occult connection between these two events. For indeed, Galileo, about to become as equally revered in science as Michelangelo was in

151

art, did seem mysteriously to have inherited a strain of that same creative talent. Whether or not Galileo's remarkable ability owed to the above coincidence, or just to the fact that for the past three centuries such precocity seemed almost genetic in the Italian population, his profound understanding of perspective drawing, especially the rendering of shades and shadows, nonetheless helped mightily to open his eyes to new revelations about nature that had escaped understanding everywhere in the world since the beginning of the human race.

A year before Galileo's birth, Giorgio Vasari, the "first art historian," had founded the *Accademia del Disegno* (Academy of Drawing) in Florence. This was intended to be an organization where painters, sculptors, and architects could meet together not as mere artisan guild members but as intellectuals, conversing about current trends in philosophy, literature, and science. Vasari wanted to establish a center where artists could keep up to date on perspective geometry and human anatomy, the sciences he believed most essential to the practice of the visual arts.

The Academy even provided for a professional geometer, an outside expert or *visitatore*, to teach perspective and chiaroscuro to less-prepared artist-members. In 1588, the twenty-four-year-old Galileo considered himself sufficiently trained in *disegno* to apply for this position. Although there is no record that he was offered the job, it was perhaps during this period that the aspiring young teacher began his lifelong friendship with the painter Lodovico Cardi (1559–1613) called Cigoli, five years older and already a member. Cigoli often lauded Galileo's knowledge of geometry, even acknowledging that in perspective drawing Galileo was his "master."[2] Galileo's increasing competence in this skill led finally, in 1613, to his own election to the prestigious *Accademia*.

Erwin Panofsky has examined their continuing correspondence, which had to do among other things with a revealing discussion of the relative merits of painting and sculpture. In 1612, Cigoli found himself embroiled in one of those endless Renaissance debates over the matter and asked his friend for support. Galileo replied that painting is surely the superior art because it imitates what is visible but not immediately tangible:

> The statue does not have its relief by virtue of being wide, long, and deep but by virtue of being light in some places and dark in others. And one should note as proof of this, that only two of its three dimensions are actually exposed to the eye: length and width (which is the *superficies* . . . that is to say, periphery or circumference). For, of the objects appearing and seen we see nothing but their *superficies*; their depth can not be perceived by the eye because our vision does

not penetrate opaque bodies. The eye then sees only length and width and never thickness. Thus, since thickness is never exposed to view, nothing but length and width can be perceived by us in a statue. We know of depth, not as a visual experience per se and absolutely but only by accident and in relation to light and darkness. And all this is present in painting no less than in sculpture. . . . . But sculpture receives lightness and darkness from Nature herself whereas painting receives it from Art.[3]

Galileo apparently cared little for the abstract vagaries of the Mannerist style as recently practiced by certain artists in his native city, preferring the volumetric, classically based painting of Raphael and advocated again by the *Accademia del Disegno*. I must also mention again that by the late sixteenth century the study of linear perspective in general and chiaroscuro in particular appealed not only to artists but ever more to professional scholars especially in Italy and Germany who otherwise had little interest in the visual arts. Numbers of highly technical perspective books were printed with this audience in mind. In Italy, prestigious mathematicians such as Federico Commandino (1509–75) and his student Guidobaldo del Monte (1545–1607) both published on the subject. Commandino was the first professional geometer to discuss linear perspective and introduce its pictorial conventions to theoretical mathematics. Guidobaldo del Monte was to become one of Galileo's strongest supporters, helping the young scientist to find his initial teaching job at the university of Pisa in 1589, and his second at the university of

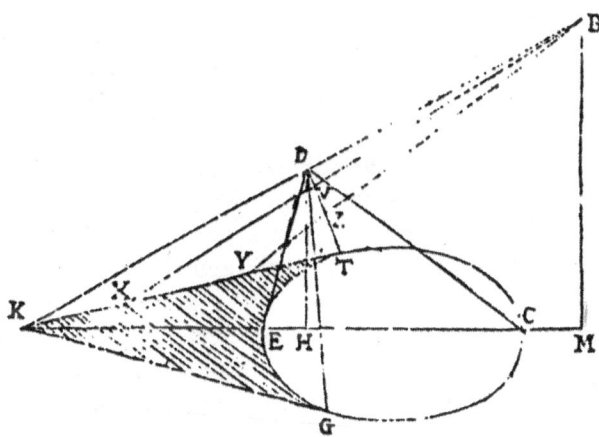

89. Guidobaldo del Monte, a page from his *Perspectivae libri sex*, Pesaro, 1600.

Padua in 1592. Guidobaldo's treatise, *Perspectivae libri sex*, published in Pesaro in 1600, contained a whole section on cast shadows and would surely have been studied by Galileo. Illustration 89 shows one of Guidobaldo's woodcut illustrations of various solids under raking light, indicating how they cast their shadows on a plane.

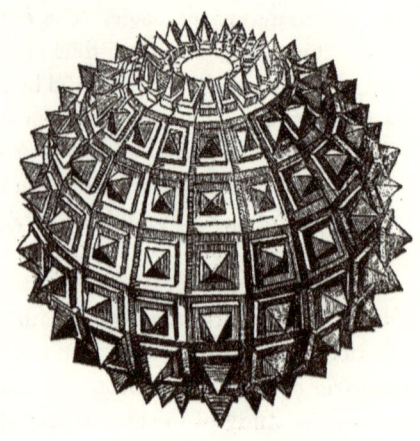

90. Daniele Barbaro, a page from his *Pratica della perspettiva*, Venice, 1568–69.

As a perspectivist, Galileo would likely have been familiar with Daniel Barbaro's *La pratica della perspettiva*, published in several editions in Venice during the late 1560s, and often consulted by members of the Florentine Accademia. Barbaro offered a number of difficult drawing exercises including how to draw spheres with raised protuberances, and how these would then receive light and cast shadows on a curving surface (ill. 90).

If Galileo were not familiar with Barbaro, he most certainly studied another similar work also entitled *La pratica di prospettiva*, by Lorenzo Sirigatti published in 1596. The latter was himself a charter member of the *Accademia* and *cavaliere* in the court of Grand Duke Ferdinand de' Medici. This handsomely published treatise consisted of two sections, the first giving standard instruction in how to project multifaceted solids and the second a series of twenty-four plates illustrating special problems of chiaroscuro, including several remarkable engravings of shaded spheres with both raised protuberances and recessed channels (ill. 91).

Let us for a moment take leave of Florence and look in on Jacobean London during the summer of 1609, where we encounter Galileo's scientific contemporary, Thomas Hariot (1560–1621), who has just procured a fascinating new instrument invented the year before in Holland, which he called a "perspective tube," and which, of course, we now call the telescope. The Dutch inventors had thought that the new device would be most useful to sailors for spotting distant ships at sea, or to military commanders for discerning far-off enemy installations, but Hariot did the novel thing of turning it on the moon. He even made a drawing of the lunar surface as he viewed it through his "perspective tube" (ill. 92).

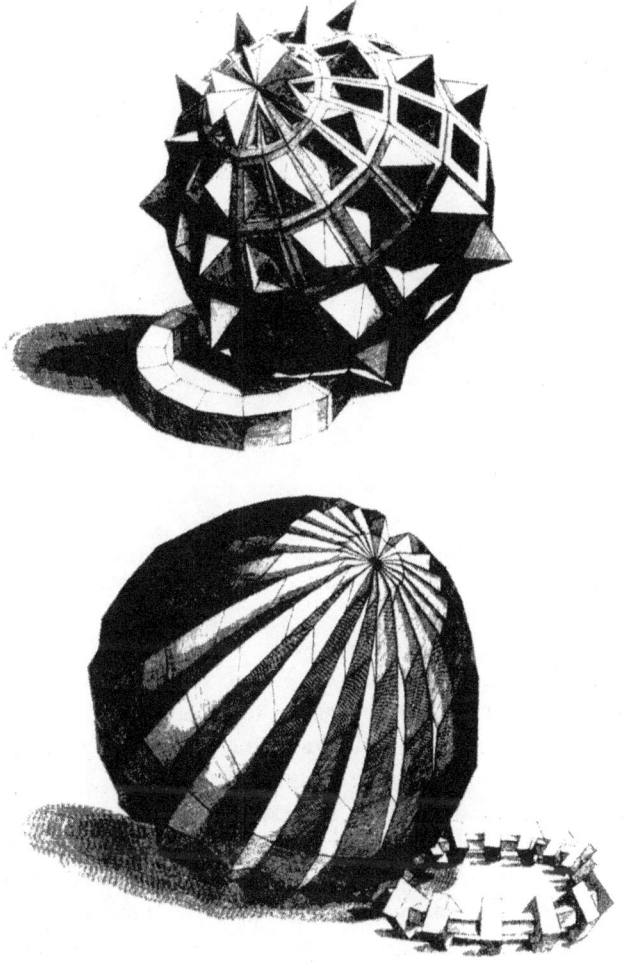

91. Lorenzo Sirigatti, a page from his *Pratica di prospettiva*, Venice, 1596.

Unfortunately, he added no explanation of the "strange spottedness" as he called the dark markings, only the Julian date and time of his observation: "1609, July 26, hor.9p.m., . . . The [first quarter] 5 dayes old." In any case, and the reason why he is so seldom recorded in books on modern astronomy, Hariot's crude sketch reveals nothing new.

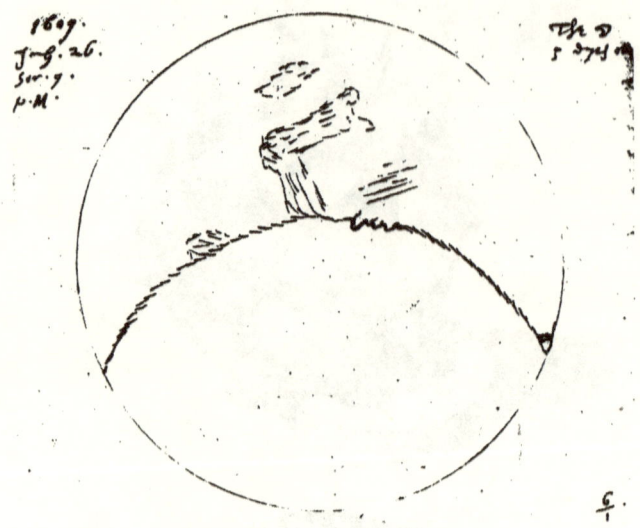

92. Thomas Hariot, drawing of the moon in first quarter. Petworth manuscripts, Leconsfield HMC, 241/ix, fol. 26, "July 26, 1609." Courtesy of Lord Egremont.

Europeans of his time still had no reason to doubt Aristotle's ancient definition of the moon as a perfect sphere, the prototypical form of all planets and stars in the cosmos. Christian doctrine added to this euphoric image by having the moon symbolize the Virgin's Immaculate Conception. "Pure as the moon" became a commonplace expression for Mary, implying that the universe, like herself, was incorruptible, that God would not have created the moon or any heavenly body in another shape. Renaissance artists, especially those serving zealous Catholic patrons, frequently depicted the Virgin standing on such a moon, as did Bartolomé Estabán Murillo (1617–82) well into the seventeenth century, especially in Spain. We see her here (ill. 93) in one of many paintings Murillo did of the subject, poised upon a ball marbled like translucent alabaster but with a highly polished, utterly smooth surface.[4]

In Thomas Hariot's England, anti-Aristotelian Francis Bacon had concluded, however, that the lunar body was not solid at all, but rather was composed of some unexplained "vapour." Hariot's own opinion about the moon's composition remains unrecorded. Nonetheless, he drew the terminator—that is, the demarcation line between the illuminated and shaded portions of the moon—with short, ragged strokes as if it fell over a roughened surface. On the upper half of the sphere, Hariot indicated the configurations of what we

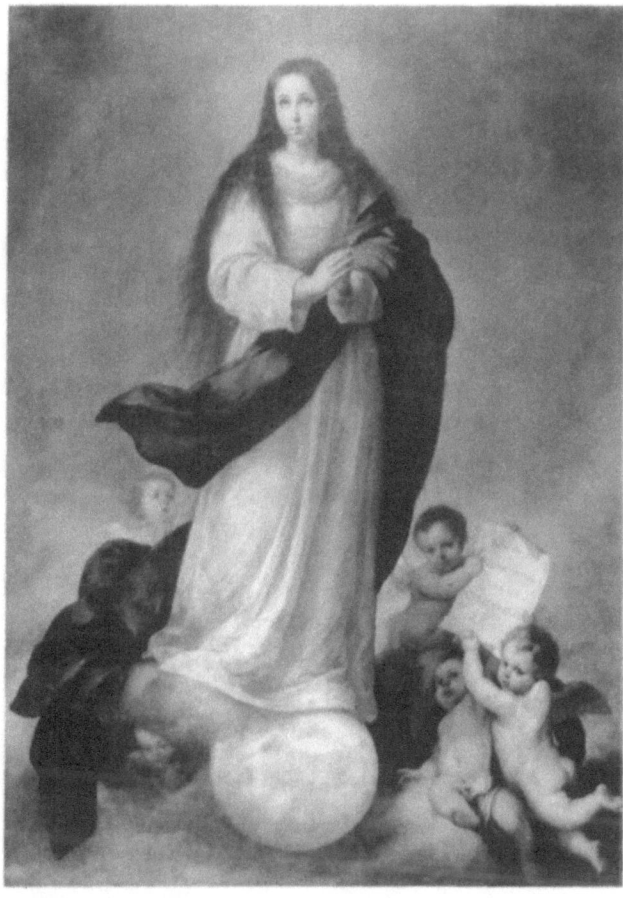

93. Bartolomé Esteban Murillo, *Immaculate Conception*, ca. 1660,
Photo © The Walters Art Museum, Baltimore.

now know as the great lunar "seas," the Maria Tranquilitatis, Crisium, and
Serenitatis, which do seem to have appeared to him as surface markings
rather than internal, vaporous discolorations. Nevertheless, he was unable to
recognize the significance of these observations. His "perspective tube" only
confirmed more or less what the ancients had always said he would see. The
"strange spottednesse" remained as mysterious to him as ever.

  Why did the Englishman miss what Galileo saw so precisely just a few
months later? Was it only because his telescope was less powerful than Galileo's? To the latter question, I answer no, because the moon through any

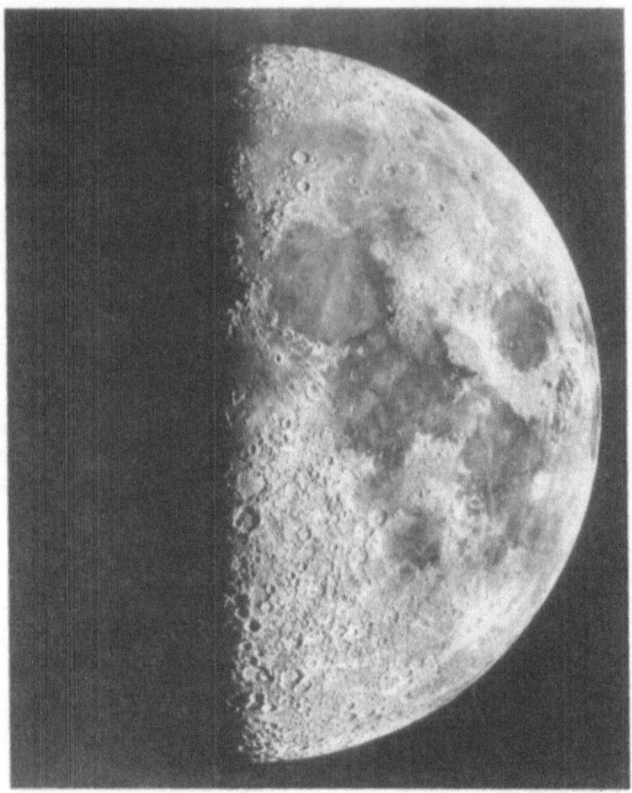

94. Photograph of the moon in first quarter, Lick Observatory, University of California, Santa Cruz. Photo © UC Regents/Lick Observatory.

"perspective tube" of the time could hardly have looked as sharp as it does in a modern Lick Observatory photograph familiar to every college astronomy student (ill. 94).

Both Galileo's and Hariot's instruments, mounted on rickety homemade stanchions, must have been difficult to focus to say the least. Moreover, as Albert Van Helden has calculated, such primitive devices had very narrow fields of view; only about a quarter of the moon could be observed at one time.[5] In sum, neither the English nor the Tuscan scientist could have seen the moon so distinctly that its true surface topography would be instantly self-evident. Besides, quite a number of such "tubes" were being produced in several centers of Europe by the end of 1609. Would not someone else also have thought to aim the instrument toward the sky? The reader might think about that while pondering the Lick Observatory photograph in illustration 94. If one knew nothing a priori about the moon's external topography, would its grayish blotches be seen immediately as shades and shadows of mountain ridges? Especially if the observer, like most people before 1610, was already certain such blotches had something to do with the moon's translucent internal composition?

In the meantime back in Padua where Galileo was teaching, the Tuscan scientist still did not know of Hariot's lunar observations. In fact, he only learned of the recent Dutch invention in May 1609. Immediately, however, he sent for instructions. With remarkable ingenuity, he applied his considerable perspective experience to the optical problems and managed by the end of the year to build a number of the instruments with magnification improved to twenty power and with the addition even of "aperture stops" to improve his image quality.[6]

Perhaps, Galileo was even reminded of Alberti's earlier demonstrations of the *camera ottica*, likewise a kind of "perspective tube." In any case, the first telescope was no more than Alberti's *camera ottica*, open-ended, extended, and enhanced by inserted magnifying lenses. His newly made-up Latin word for the instrument was *perspicillum*, singular of the same neuter-plural term for "magnifying eyeglasses." Van Helden has translated this new word use as "spy glass."[7] In his vernacular longhand notes, Galileo referred to it simply as *il cannone* or "the tube."[8]

Investigation has shown that Galileo's recordings of the moon's phases date from November 30 through December 18, 1609, and again on January 19, 1610.[9] Since his observations during these brief periods could only be affirmed when the moon appeared in partial shadow, his viewing nights were quite limited, nor would all be conveniently clear and cloud free.

Perhaps Galileo made some illustrations right there on the spot as he stared at the moon from atop the San Giorgio Maggiore campanile in Venice. Although none of these have survived, we are in possession of seven finished sepia studies, which I believe were done later, based on his on-the-spot sketches. These small finished wash drawings, four of the waxing and three of the waning moon, are still preserved on two sides of a sheet of artist's watercolor paper in the Biblioteca Nazionale in Florence. All were certainly done by someone well practiced in the manipulation of ink washes, especially the rendering of chiaroscuro effects. They are by an experienced artist, and we have no reason to believe by anyone other than Galileo himself (ill. 95).

Galileo no doubt prepared these washes as models for the engraver who would illustrate his book, *Sidereus nuncius* (Messenger from the Stars), which

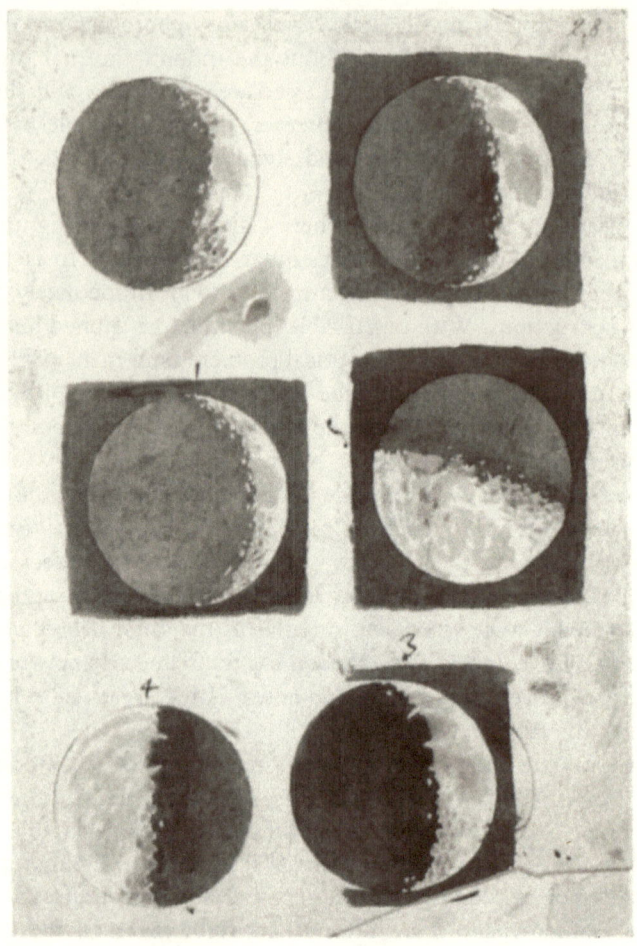

95. Galileo, wash drawings of the moon (page 1), 1609. Ms. Gal.
48, fol. 28r. Biblioteca Nazionale Centrale, Florence.

he rushed into publication in March 1610, barely four months since he began
looking at the skies through his homemade telescope. Only five engravings
of the moon's phases were printed in *Sidereus nuncius*, none exactly replicating
the wash drawings.[10] Illustration 96 indicates how two of these appeared in
Galileo's book.

Our Lick Observatory photograph, illustration 94, shows the same waxing moon as depicted on the page left above in *Sidereus nuncius*. Galileo's ac-

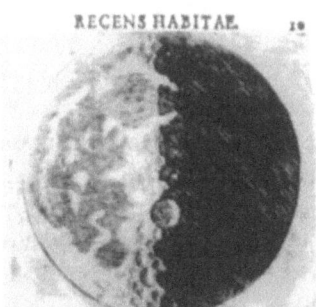

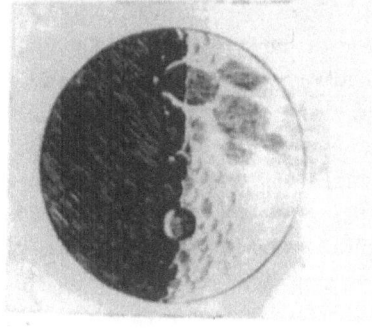

96. Galileo, facing pages from his *Sidereus nuncius*, 1610. Courtesy of Jay M. Pasachoff, Williamstown, Massachusetts.

companying matter-of-fact textual description of these engravings belies both his own excitement and the stupendous impression they made on an unsuspecting world:

> [I] have been led to the conclusion that . . . the surface of the Moon is not smooth, even, and perfectly spherical, as the great crowd of philosophers have believed about this and other heavenly bodies, but, on the contrary, to be uneven, rough, and crowded with depressions and bulges. And it is like the face of the Earth itself, which is marked here and there with chains of mountains and depths of valleys.[11]

As stated, the illustrations in *Sidereus nuncius* are not exact copies of any of the wash drawings. It would seem that Galileo furnished the latter only as guides to the engraver, who was apparently asked to emphasize the more spectacular features of the moon's surface. He even permitted the engraver a certain artistic license to exaggerate the size of that particularly dark, deep crater we see lying just below center along the terminator in both engravings in illustration 96. This is Albategnius, and Galileo wished to compare its steep sides to the high mountains on Earth surrounding the region of Bohemia. Thus he bade his engraver to render it large, to dramatize that the moon is covered all over with such rugged depressions. We should also bear in mind that the engraver would probably not have looked through the telescope himself, but depended solely on the astronomer's drawings and, no doubt, on Galileo's rather excited verbal descriptions.

Galileo's original wash drawings reveal a much more "painterly" lunar surface than do the published engravings. Most modern historians have talked about only the latter, which by virtue of their metallic, linear technique, make Galileo's moon look like the arid and lifeless body our modern astronauts discovered it to be. His wash renderings, on the other hand, show that he still regarded the moon somewhat in the old medieval "watery" spirit. With the deft brushstrokes of a practiced watercolorist, he laid on a half dozen different grades of washes, imparting to his images an attractive soft and luminescent quality. Remarkable indeed was Galileo's command of the Baroque painter's convention for contrasting lighted surfaces, and his ability to marshal dark and light washes to increase their mutual intensities. In the upper left of the sheet of sepia drawings in illustration 95 we see how he set down a little practice patch of dark and light washes surrounding a white area, probably to help his engraver realize the form of the lunar crater as it took shape in the waxing light. With artistic economy worthy of Tiepolo, Galileo indicated the concave hollow with a single stroke of dark, leaving a sliver of exposed white paper to represent the crater's glowing brim.

Is it preposterous to claim that these simple yet highly professional paintings belong as much to the history of art as they do to the history of science? Indeed, we have considerable evidence testimony of Galileo's skill as a draftsman and of his interest in the visual arts.[12] In the true spirit of the Florentine *Accademia*, Galileo seems to have engaged in *disegno* not for the sake of self-expression but rather to discipline his eye and hand for science. And yet he has at the same time in these chiaroscuro washes anticipated the independent landscape in the history of art. His almost impressionistic technique for rendering fleeting light effects anticipates Constable and Turner, and perhaps even Monet. One needs only to read on in *Sidereus nuncius* to appreciate his wonder, as well as his rational understanding as he gazed upon the transient moonscape, noticing it was covered with small spots having

their dark part on the side toward the Sun, while on the side opposite the Sun they are crowned with brighter borders like shining ridges. And we have an almost entirely similar sight on Earth, around sunrise, when the valleys are not yet bathed in light but the surrounding mountains facing the Sun are already seen shining with light. And just as the shadows in the earthly valleys are diminished as the Sun climbs higher, so these lunar spots lose their darkness as the luminous part grows. Not only are the boundaries between light and dark on the Moon perceived to be uneven and sinuous, but, what causes even greater wonder, is that many bright points appear within the dark part of the Moon, entirely separated and removed from the illuminated region and located no small distance from it. Gradually after a small period of time, these are increased in size and brightness. Indeed, after 2 or 3 hours they are joined with the rest of the bright part, which has now become larger. In the meantime, more and more bright points light up, as if they are sprouting, in the dark part grow, and are connected at length with that bright surface as it extends farther in that direction. . . . Now on Earth, before sunrise, aren't the peaks of the highest mountains illuminated by the Sun's rays while shadows still cover the plain? Doesn't light grow, after a little while, until the middle and larger parts of the same mountains are illuminated, and finally, when the Sun has risen, aren't the illuminations of plains and hills joined together?[13]

Did ever a Baroque painter express the new secular spirit of landscape art better than this? Was ever an artist's eye better prepared to recognize the universal geometrical principles of perspective optics and chiaroscuro even at work on the moon? Moreover, after thus having marveled at the picturesque lunar terrain, Galileo quickly reverted to his scientific self and made two other amazing perspective-related discoveries. The first was when he noticed that some of the lunar peaks were tipped with light within the shadow side even as the terminator boundary lay a long way off. At the same time, he was able to convert this phenomenon into a geometric diagram for solving a shadow-casting problem such as he may have recalled from Guidobaldo del Monte.

Illustration 97 illustrates another manuscript page that Galileo prepared for *Sidereus nuncius*. On it he drew a circle representing the moon, divided by the terminator, which he marked *cef*. The sun's shadow-casting light rays he indicated by the tangent line *dcg*. With particular ingenuity, considering that his primitive telescope had no crosshair sighting device, he was able to estimate the real distance of the lighted lunar mountain peak to the terminator as being about one-twentieth—line *dc* here in the diagram—of the moon's whole diameter. This distance, more or less comparable to line *DK* in Guidobaldo del Monte's cone/shadow diagram illustration 89, then allowed him to triangulate the mountain's height, almost as if he were standing on the moon itself holding Brunelleschi's staff. Since the moon's diameter was known to be two-sevenths of Earth's own diameter, or about two thousand miles, Galileo's

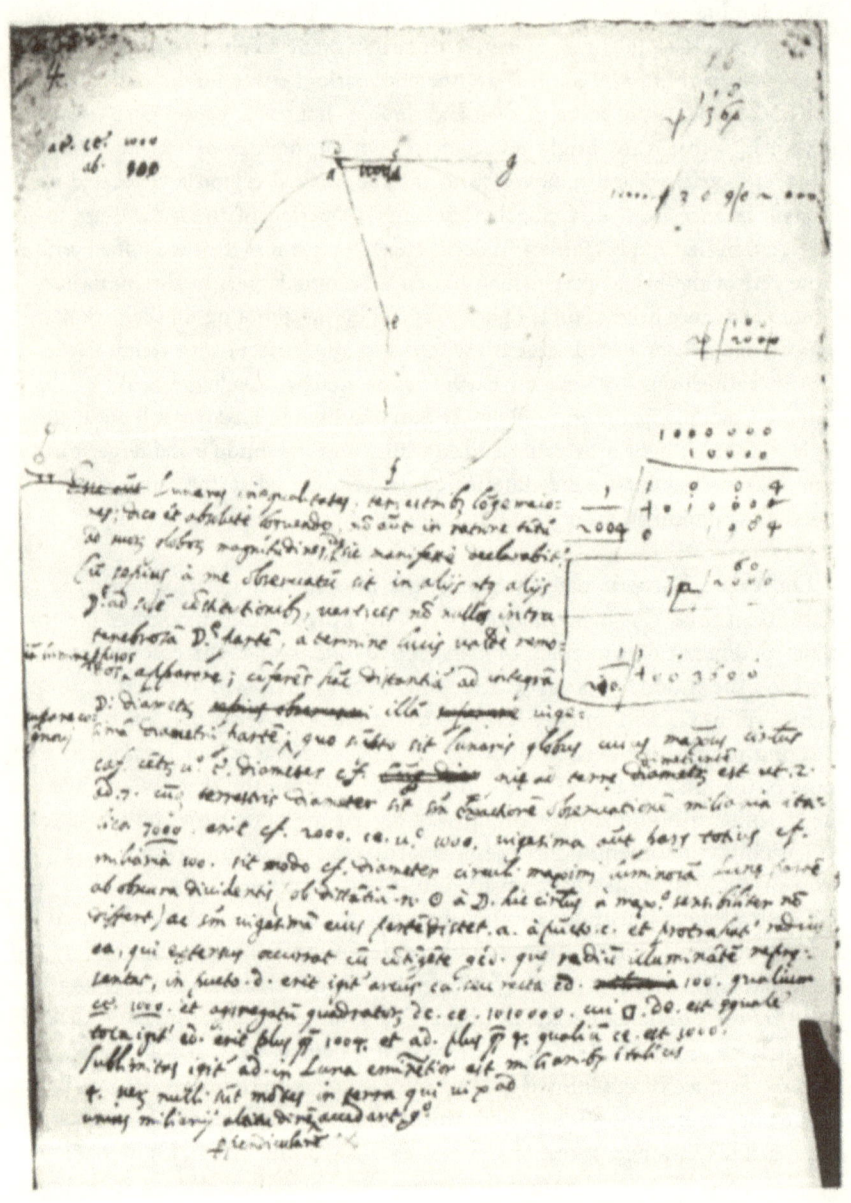

97. Galileo, manuscript page showing the triangulated heights of the moon mountains. Ms. Gal. 48, fol. 16r. Biblioteca Nazionale Centrale, Florence.

triangle *ced*, with *ce* equaling one thousand miles, and *cd* one hundred, revealed by Pythagorean calculation that *da*, the mountain's height on center from its base, reached more than four miles into the lunar sky. By applying a problem well known to students of Renaissance perspective, Galileo added yet another fact to his already wondrous revelations, that the mountains on the moon were more spectacular than the Alps here on Earth.

Galileo's telescopic observations of the moon announced in *Sidereus nuncius* opened the eyes of Renaissance Europeans to a celestial reality they had never before imagined. If Thomas Hariot's Britain still lingered in the Middle Ages, Galileo offered that insulated land a crash course in Italian ways of seeing. Suddenly everywhere in Britain, amateur as well as professional philosophers were able to see the same sunlit mountains and shadowed valleys on the moon just as Galileo had described, whatever the quality of their own telescopes. The lunar landscape as well as the "perspective glasse" soon became metaphors in the writings of John Donne and many later British poets.

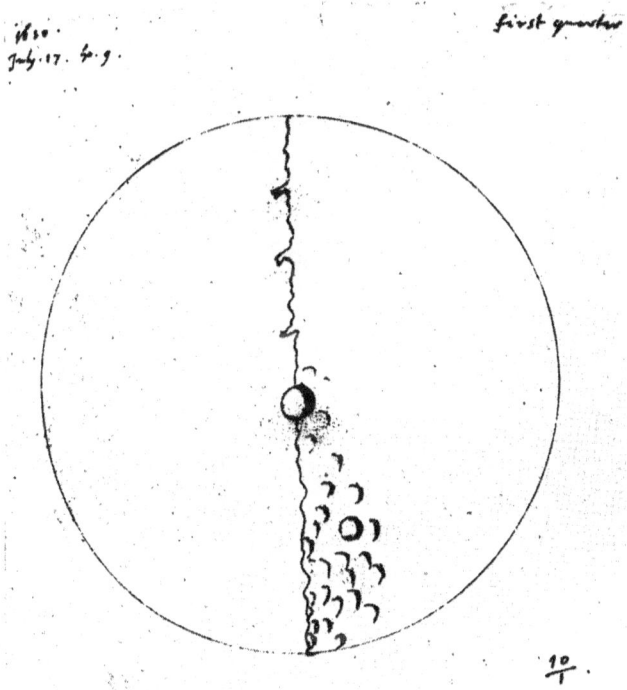

98. Thomas Hariot, drawing of the moon in first quarter. Petworth Manuscripts, Leconsfield HMC, 241/ix, fol. 20, "July 17, 1610." Courtesy of Lord Egremont.

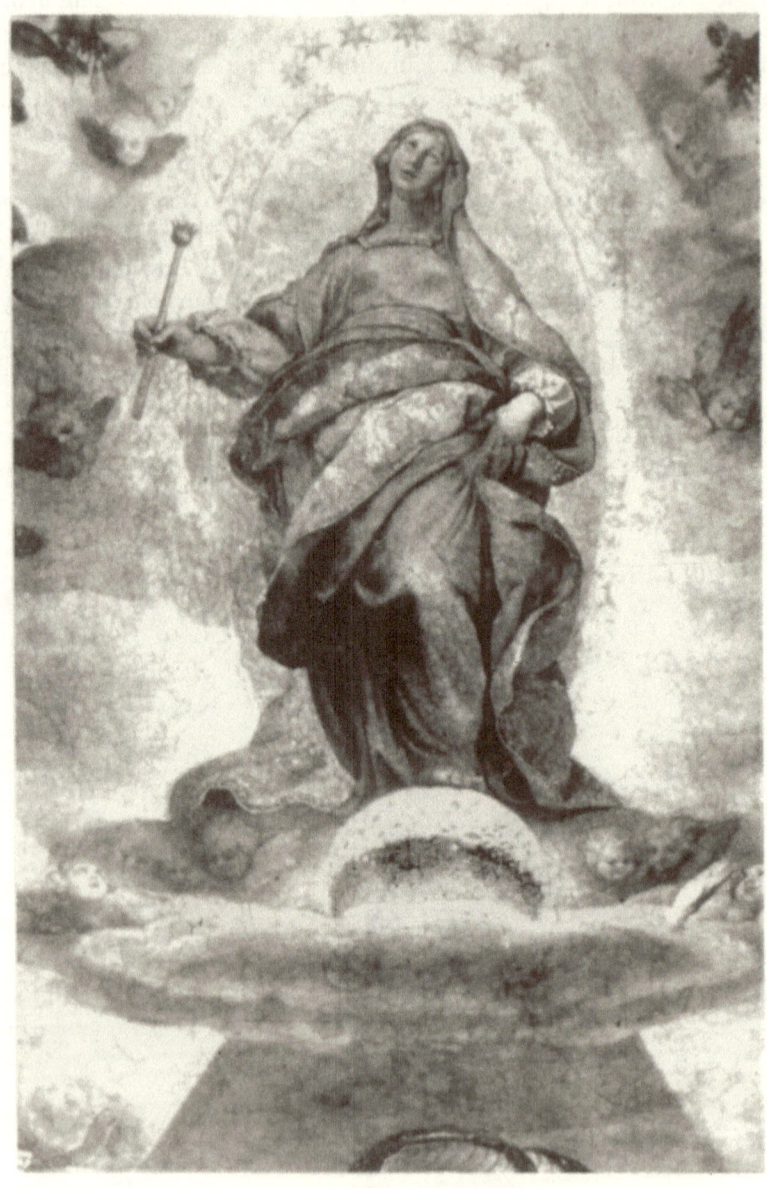

99. Cigoli, *Woman of the Apocalypse*, Pauline Chapel, Santa Maria Maggiore, Rome, 1612. Courtesy of Miles Chappell, Williamsburg, Virginia.

Even Hariot, once he had read *Sidereus nuncius*, finally "saw" the shaded craters that had eluded him a year before. In July 1610, four months after *Sidereus nuncius* was published, Hariot drew yet another lunar picture (ill. 98).[14]

Again, there is no written comment, but the Englishman did sketch the moon's concavities in pen-stroke circles and half circles, even exaggerating Albategnius in imitation of the *Sidereus nuncius* engraver's drawing. It is a curious fact, if only a coincidence, that in 1611, hardly a year after England received Galileo's stunning announcement, Inigo Jones (1573–1652), the first Englishman to have talent and training in the conventions of Italian perspective drawing, was appointed Surveyor General to the Prince of Wales, and Sebastiano Serlio's *Treatise on Architecture*, the most widely read textbook on the neoclassical style, including its special section on linear perspective, was translated into English, also in 1611. Both events, following immediately upon the news of Galileo's telescopic discoveries, signaled the arrival finally of the full-blown Italian Renaissance to the British Isles.

There still remained, of course, some recalcitrant souls who so firmly believed the moon was "pure" that they could not be persuaded to look through Galileo's telescope. The Roman Catholic Church, however, was quick to co-opt the new discovery. In 1612, Galileo's friend Cigoli the painter was commissioned to fresco the domed ceiling of the Pauline Chapel in the Basilica of Santa Maria Maggiore in Rome. The artist was permitted to depict there the Virgin Mary standing on a crater-pocked Moon, no doubt inspired by one of Galileo's original drawings (ill. 99).[15]

To this day Cigoli's painting is officially and prudently called the *Woman of the Apocalypse* and not the *Immacolata*. By this admission in such a sacred place, the Church tacitly acknowledges that Galileo was not altogether wrong about at least some of the heavens looking just like Earth.

It is worth noting in conclusion, however, that as astronomers after Galileo demanded to see ever more distant planets and stars, the perspective tube had to be extended longer and longer. Finally, no less than Isaac Newton realized that another optical component must be added to increase the instrument's power to see deeper into the ever more "enigmatic" heavens. "Alberti's window" needed further enhancement by the supplementation of Brunelleschi's mirror.

Postface
Post perspectivam

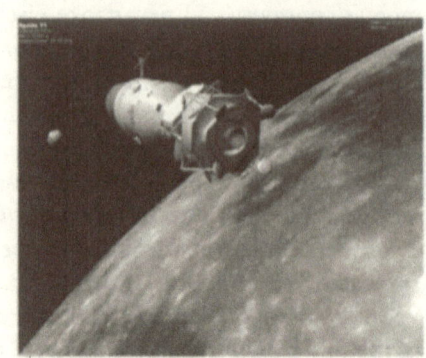

100. Simulation illustration of Apollo II spacecraft circling the moon. Image: NASA.

> We see the task of science rising before us as an incessant struggle toward a goal which will never be reached, because by its very nature it is unreachable. It is of a metaphysical character, and, as such, is always again and again beyond our achievement.
>
> Max Planck, 1931

Once Galileo's telescope had revealed the moon to be just another terra incognita with a landscape similar to much of what was being explored in the new American continents at that very moment, it was inevitable that voyaging to the moon would likewise seize the Western imagination.[1] While much fantasized about by literary dreamers such as Daniel Defoe, Jules Verne, and H. G. Wells, its factual accomplishment by Neil Armstrong in 1969 could never have happened at all without the Renaissance invention of pictorial linear perspective and its later derived science of descriptive geometry (ill. 100).

Indeed, linear perspective may have been more important to the history of modern technology than to art and science. With this remarkable tool it was now possible to plan complex machines by means of exacting pictures to scale, thus allowing each connecting part to be previewed and tested from every aspect on two-dimensional paper before being manufactured into irrevocable

168

three-dimensional fabric. Even modern computer modeling is based on the rules of linear perspective.

Prior to Brunelleschi's demonstrations, mechanical apparatuses of whatever sort were never constructed from scale drawings. Although pictures were sometimes used by medieval artisan engineers, they were only to suggest the general purpose of a machine to be constructed, as was the quaint drawing in illustration 101 of a water pump from a fourteenth-century Islamic manuscript. Any skilled craftsman of that time who already knew how to build such devices would only glance at such an image merely as a reminder of his job. In any case, it's obvious that this picture is hardly an accurate diagram-to-scale from which a three-dimensional working model could be fabricated.

Look now at the two following "postperspective" illustrations from mid-fifteenth century Italy. The first, illustration 102, is a drawing of a suction pump by the Sienese engineer called Taccola (1381–ca. 1453). Although only a modest artist, he was still fully aware of Brunelleschi's optical rules and how to maintain proper scale in his pictures so that whatever he was designing could be easily translated into an actual working model. But there's a flaw evident in this drawing. As the crank at the top would turn, the rope that pulls the piston up and down must oscillate back and forth with every rotation, causing the piston to rub each side of the circular wellhead until it is eventually bent into the shape of an oval, and thus cause the pump to lose suction and become dysfunctional.

What's interesting is that this pump never needed to be built in order to prove, at expensive cost, that it would indeed quickly fail. Taccola's immediate successor, Francesco di Giorgio Martini (1439–1501), engineer, architect, and painter, realized the problem instantly, simply from studying this earlier drawing. Here in illustration 103 is his pictorial correction.

Without even having to reconstruct a model of the old pump in order to test its oscillating action, he was able to redesign it with an ingenious improvement. Notice that the crank in his drawing now has a rolling slip ring around it, and the piston rod has a loop in the top in which the slip ring can roll back and forth so the piston only goes straight up and down, and never wobbles, thus causing no damaging friction.[2]

These simple drawings, dependent as they were on the draftsmen's knowledge of linear perspective to scale, indicate graphically what this unique Renaissance artistic technique bequeathed to modern technology. To repeat, linear perspective drawing to scale made it possible to conceive, improve, and modify the most complicated machinery without having to waste time and money building and testing expensive three-dimensional models. No

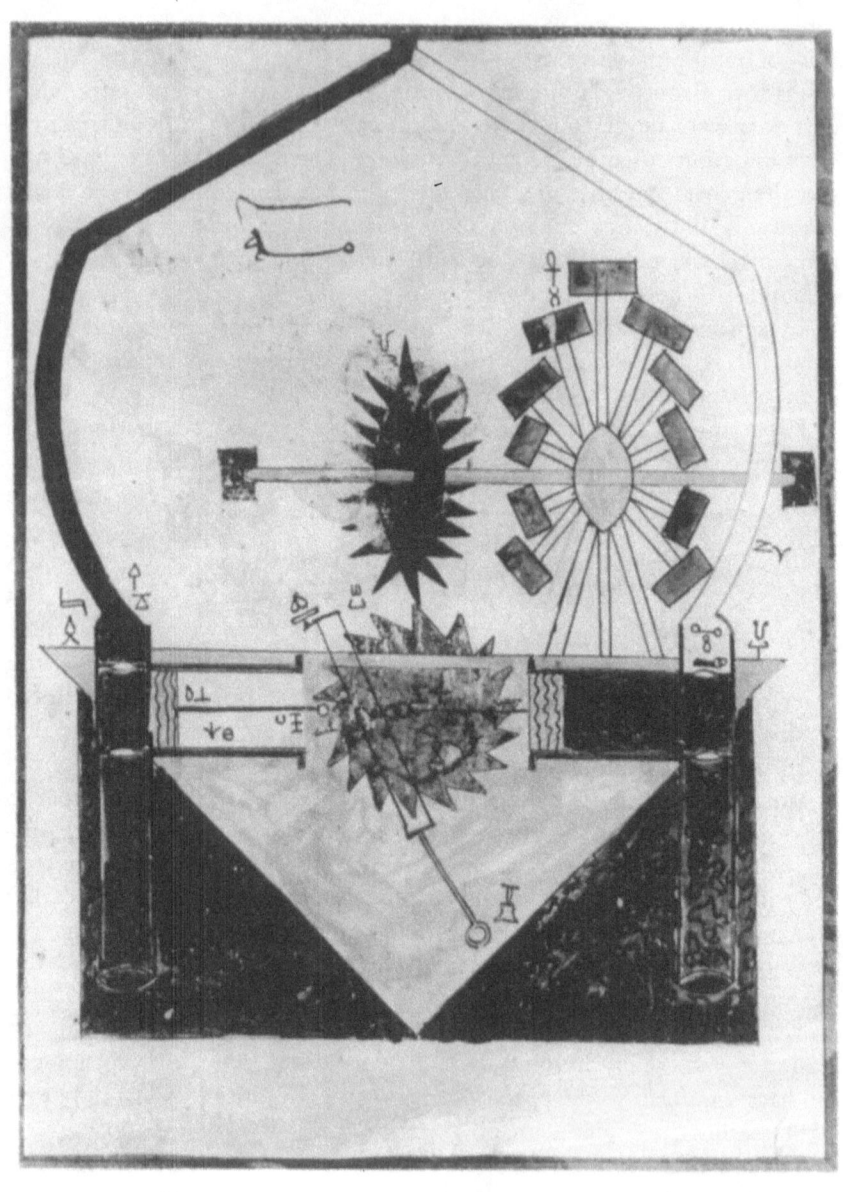

101. Ibn al Jazari, a page from his *Compendium of the Theory and Practice of the Mechanical Arts*, ca. 1310. Denman Waldo Ross Collection, Museum of Fine Arts, Boston.

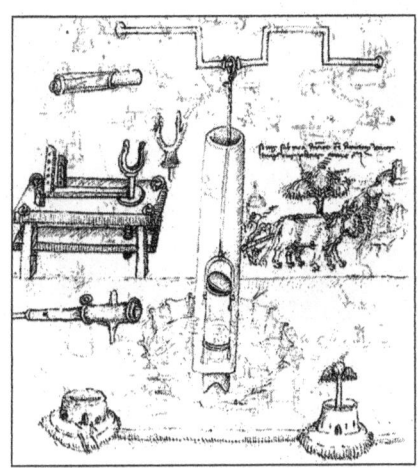

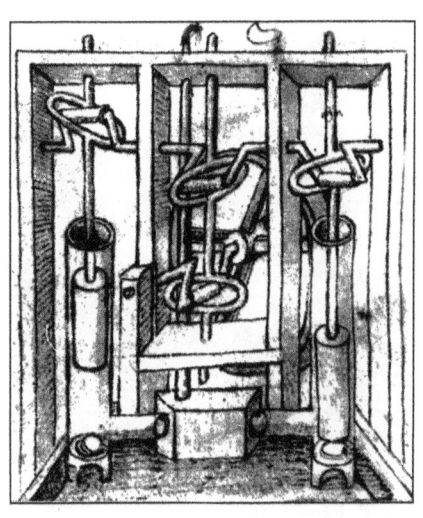

102. Taccola, a drawing from his *De machinis*, ca. 1443. Bayerische Staatsbibliothek München / Clm 197, II, fol. 82v.

103. Francesco di Giorgio Martini, detail from a page in his *Trattato*, ca. 1470. fol. 42v. Florence, Biblioteca Laurenziana. By permission of the Ministero per i Beni e le Attività Culturali.

rocket ship to the moon could ever have been invented, let alone be built and function, without the humble heritage of Renaissance linear perspective.

Finally, what of Alberti's window in this "postperspectival age"?[3] Let me briefly allude to the recent technology of "digital image processing" that has almost entirely replaced conventional photography in astronomical observatories, especially for recording the most distant galaxies and quasars. This ability is the result of a remarkable postage-stamp-sized silicon chip known as the CCD, or "charged-coupled device," which can be attached to a telescope to make it into a camera, and which acts like a photographic plate. Rather than being coated with light-sensitive emulsion, its surface is composed of millions of tiny electronic "pixels" arranged in a rectangular grid.[4]

Have we here the ultimate heir to Alberti's window? Just as the Renaissance artist transferred his earthly subject, square by square as seen through the gridded veil, onto his smaller-scaled picture surface, so the CCD collects cosmic light through the telescope, pixel by pixel, and then converts each photon impulse into a digitized electronic signal and sends it to a television monitor. Celestial bodies are perceived and revealed against the deep-sky background by the luminosity fluctuations and changes in brightness recorded by the individual pixels across the CCD grid. These digitized encodings can then be translated

into computerized colors. Since the
actual hues of the celestial subject remain invisible, however, the image
processor (as the CCD artist is now
professionally called), even though
equipped with spectrum sensors and
chromatic filters, must still make decisions based on aesthetic preference
just as do modern artists—or one
might even say like medieval artists
trying to comprehend the ineffable
colors of the heavenly empyrean.

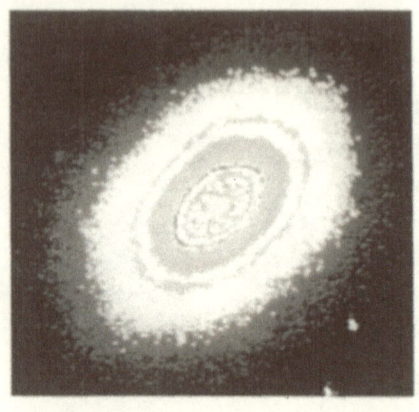

Illustration 104 is a good example
of the modern processor's art. It is a
CCD image of one of the gigantic
elliptical Virgo galaxies some sixty

104. CCD image of the Virgo galaxies.
Photo courtesy of Jay M. Pasachoff,
Williamstown, Massachusetts.

million light years distant from Earth. In the image, notice the rows of pixel
dots, each displaying a single point of varying luminosity as received and
recorded electronically from the object in focus. In the purely flat format of
this electronic technique, the illusion of depth can only be signified by a
value contrast between hues. Just as modern artists have tended to reject
Renaissance-style black and white chiaroscuro, preferring instead to model
their forms by means of contrasting complementary colors, so the astronomer processor of this image has chosen warm yellow and orange to show
what his digital information indicates must be the brightest and perhaps most
active part of the object, and cool blue and green to indicate the part that is
perhaps behind or receding away.

Illustration 105 is another remarkable CCD picture of what is known in
astronomy as "gravitational lensing" in which light rays coming from galaxies
billions of light years away actually bend as they pass the heavy gravitational
fields of intervening stars, thus creating multiple impressions of themselves as
they reach the CCD chip. This is quasar G2237–0305 as recently recorded by
the orbiting Hubble space telescope, showing the same quasar's repeated image four times as its light curves around a nearer galaxy in the center. Once
again Nature signifies one of her deepest mysteries with an esoteric sign, the
primordial quincunx, which the processor has interpreted in stark white em-
blazoned against a deep purple glimmer spreading in the black void.[5] One can
almost share the excitement of the awestruck artist that what was being pro-
cessed through the pixel "window" might well be the very light that left
God's eye at dawn on the first day of Genesis.

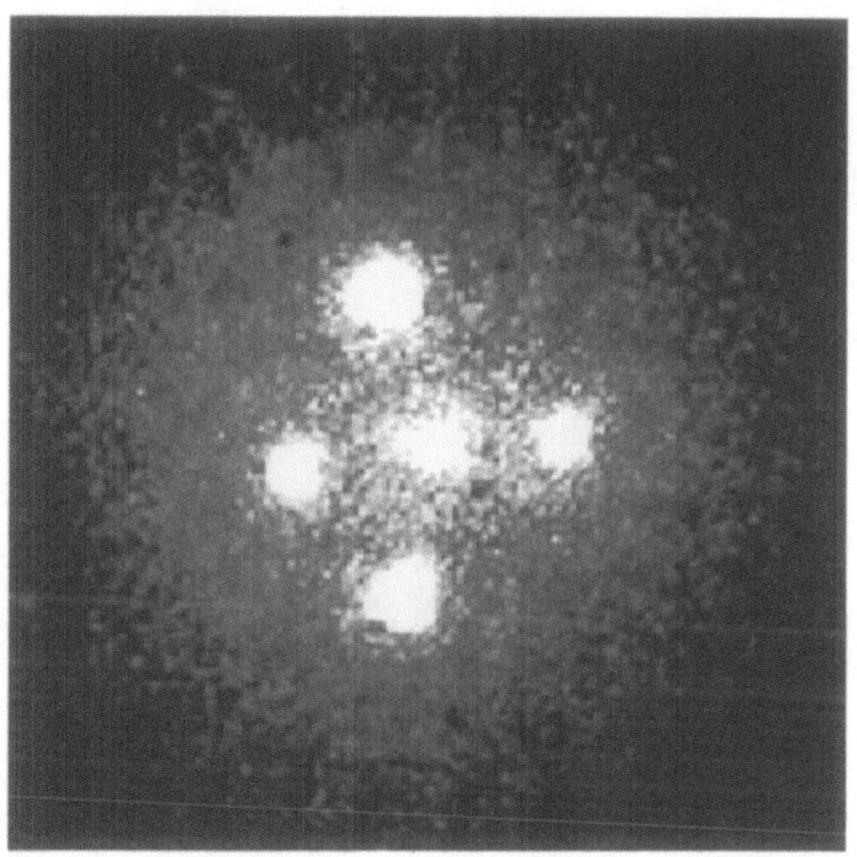

105. The "Einstein Cross," Quasar G2237+0305. Photo: NASA.

Astronomers have dubbed this attractive figure the "Einstein Cross," an appropriate symbol that fuses religious, artistic, and scientific concerns that have been the subject of this book. Yet the question remains: Is the image only a secular creation of Alberti's digitized window, or is it a spiritual reflection from the vestige of Brunelleschi's mirror? Whichever, it's still the result of art and science working together in the never-ending quest to picture the "reality" of our ineffable universe.

Notes

Preface

1. Perspective diagram from Vignola.

Introduction: Picturing the Mind's Eye

1. Except in strange cases among autistic children, like the remarkable Nadia. See Golomb: 262–71.

2. See J. Gibson on how we perceive perspective in nature even if we don't naturally depict it. During World War II the author developed his theories by observing how airplane pilots watch for clues of diminishing perspectival distance as they descended rapidly to land on often darkened fields. See E. Gibson and Pick for how much of this perceptual information is "hardwired" in our brains even as tiny babies.

3. On children's art in general, see the many examples and analyses of how they perceive "perspective" by Winner, Kellogg, and Gardner, as well as Golomb.

4. Panofsky (3).

5. See Grafton (2) for a rich discussion of how Latin reigned as the lingua franca among the educated classes all over Western Europe and their dominions right up through the eighteenth century.

6. In all further references to Alberti's theories in this book, I will be citing from the recently published English translation of the 1540 edition of *De Pictura* by Rocco Sinisgalli; see Alberti (3). Although Sinisgalli's English is not as elegant of that of Cecil Grayson, whose own translation—see Alberti (2)—has until now been the most quoted in the relevant literature, Sinisgalli's is more literal, and in many instances more exact than Grayson's. For an explanation of his differences with Grayson, see especially Sinisgalli's introduction in Alberti (3): 25–66.

2. Crisis in Christendom

1. For a beautifully illustrated history of cartography throughout the world, see Woodward and Harley: (1), (2).
2. Bacon (3), II: 581.
3. Translated in Bacon (3), I: 232–34; the original Latin is in Bacon (1), I: 210–11.
4. See Edgerton (5): 47–88.
5. These were called *Mirakelspielen* in Germany, *laude* in Italy, *mystère* in France, and *autos* or *pastorales* in Spain.
6. Concerning the prevalence of Vitruvian manuscripts in northern Europe well before the Renaissance, see Schuler.
7. Pollitt: 236–47.

3. And God Said, "Let There Be Light!"

1. See Bacon (1), II: 405–552; Pecham; Lindberg (2); and Witelo.
2. For an English translation and commentary on Alhacen's work and influence, see Smith.
3. The *Thesaurus Linguae Latinae* X.I, fasc. xi: 1736, merely mentions *perspectiva*, without further definition, as "a term of the Middle Ages." However, the word still did not make the Du Cange Dictionary of Medieval Latin, nor is it to be found in the recent *Novum Glossarium Mediae Latinatis*, covering Latin usage between 800 and 1200 AD. However, on page 730 of the latter, several definitions are given of the root verb, *perspicio*, but none of these have to do with either the visual arts or the science of vision. It would seem, therefore, that not until the thirteenth century did *perspectiva* become the common Latin term for optics.
4. Concerning the study of *perspectiva* during the Middle Ages in Europe, see Lindberg (2) and Tachau.
5. Pecham: 34.
6. Schwarz.
7. Pecham: 120–21.
8. See Bialostocki; Pendergrast.
9. See Crombie: 99–104.
10. Pfeiffer: 84–85.
11. Edgerton (2): 87–88. For the original Italian in context, see Alighieri: 261–62.
12. As translated by Lindberg (3): lxiii. A slightly different version is offered in Bacon (3), II: 489–90.

4. Fra Antonino

1. See Lindberg (1); also Steinberg and Edgerton: 52, notes 19–20.
2. D. L. Clark.
3. Note particularly the passage in Antonine I, Tit. II, cap. 6, cols. 88–93, where he speaks of "a sensible thing which is seen, multiplies its species (*multiplicat speciem suam*), that is its likeness at every point according to the visual power."
4. Antonine IV, Tit. 9, cap. i, col. 465–66.

5. Ibid. I, Tit. 3, cap. iii, col. 117.

6. Ibid. Antonino used the similar adverb, *directe*, clearly in the *axis perpendicularis* sense later in the *Summa* when he speaks of why the land is warmest at noon when the sun's rays strike it most directly. IV, Tit. 9, cap. i, col. 467.

7. Antonine I, Tit. 3, cap. iii, col. 117.

8. Ibid.: col. 110. I thank Professor Rocco Sinisgalli for calling this pithy passage to my attention.

9. Ibid.: col. 122.

10. Pecham: 249, note 93.

11. Antonine I, Tit. 3, cap. iii, col. 118.

12. *The New American Bible* first published in 1970 by the Bishops' Committee of the Confraternity of Christian Doctrine, Washington, D.C., is the result of a twenty-five-year project "to create a new translation of the Scriptures from the original languages or from the oldest extant form in which the texts exist." It was authorized by Pope Pius XII in his 1943 Encyclical *Divino afflante Spiritus*, and the Decree of the Second Vatican Council.

13. Antonine I, Tit. 3, Cap. vi, col. 144: *quod est imago, quasi sententia sive locutio facta sub imagine, idest sub similitudine. . . . Sicut enim ipsi vident Deum, sive divina mysteria sub imaginibus et similitudinibus rerum sensibilium.*

14. Ibid.: col. 137.

15. Goldberg.

16. Edgerton (4).

17. Hartt: 227–28 and passim.

18. For a more thorough examination of this painting in the same light, see Steinberg and Edgerton: 45–53.

19. Bacon (3) I: 131–36.

20. Ibid. II: 471.

21. Ghiberti (2) II: 94–95.

5. When Did *Perspectiva naturalis* Become *Perspectiva artificialis*?

1. Tanturli: 125.

2. See Baron II: 548, note 15 for the context of the letter; the full Italian text is given in Wesselofsky: 336f; see also the *Enciclopedia dantesca* II: 552–53 for a further note on the life of Domenico da Prato.

3. Kemp: 9. James Elkins also claims, likewise erroneously, that this 1413 document is a "recently discovered mention of panels by Brunelleschi"; Elkins: 8.

4. Actually, Domenico in later years became a severe critic of Brunelleschi's architecture.

5. See the four-page listing of various literary citations for *prospettiva* in the *Grande Dizionario* XIV: 710–13, not one of which refers to the visual arts before the late fifteenth century. Also the *Thesaurus Linguae Latinae* in its similar long listings of all forms of the verbs *perspicere* and *prospicere* offers no meanings that could be construed as having to do with the art of depiction.

6. Pliny XXXV: 81–84; the Roman author uses only the words *de mensuris*, "according to the measurements," but some modern authors still translate this classical

Latin phrase as "perspective" (Jex-Blake: 121–23). Ghiberti's version has been edited by
O. Morisani in Ghiberti (1): 21.

7. Averlino II, fols. 178v–179r: È veramente da questo modo credo che Pippo di Ser
Brunellesco trovasse questa prospettiva, la quale per altri tempi non s'era usata.

8. Saalman: 42–43: "Cosj ancora in que tempo e misse innanzi et innato luj proprio
quello che dipintorj oggi dicono prospettiva, perche ella e una parte di quella scienza
che e in effetto porre bene e con ragione le diminuizionj e acrescimenti, che appaiono
agli occhi degli huomini delle cose di lungi e d'apresso."

9. See, for examples, Krautheimer: 229–53; and Kemp: 344–45.

10. Antonine II, Tit. 3, cap. vi: col. 132.

6. Brunelleschi's Mirror

1. Saalman: 42–44. Copyright 2008 by The Pennsylvania State University.
Reproduced by permission of the publisher.

2. Edgerton (2): 124–52.

3. Villani: book 12, chapter XVIII: 456.

4. Antonine IV, Tit. 16, cap i, col. 1265.

5. Krautheimer I: 34.

6. Ibid.: 159.

7. Parronchi: 313–49.

8. Antonine IV, Tit. 16, cap i, col. 1265.

9. On the availability and display of mirrors in fifteenth-century Italy as well as
their types, see Thornton: 234–41. On their manufacture and quality, see
Melchior-Bonnet and especially Schechner. The Boston Museum of Fine Arts possesses
a flat silver-coated metal mirror of fifteenth-century Florentine provenance (accession
number 55.621). It is mounted in a decorated rectangular frame of later style and appears
to be about ten by seven inches in size.

10. Still another factor, if at the time only unconsciously perceived, may explain why
Brunelleschi found his little hole so effective in enforcing his illusion—what the
philosopher of science Michael Polanyi has called "subsidiary awareness." The hole in
effect would so focus the viewer's eye directly upon the image that the edges of the
mirror would become quite occluded, causing the viewer to lose sense of the surface on
which the image appears. This in turn makes the image seem to float free of the
surface, as in a hologram, and look startlingly three dimensional.

11. In their controversial 2001 book Secret Knowledge, British artist David Hockney and
University of Arizona physicist Charles Falco attempted to prove that Renaissance painters
"secretly cheated" by using parabolic mirrors to project images onto their picture surfaces
that they then traced, rather than drawing them freehand. See Schechner for a convincing
rebuttal of their claim. Although fifteenth-century Flemish artists no doubt did copy from
images reflected in convex (the reverse side of parabolic) mirrors, no documented evidence
has been found to sustain the author's assertion that they employed or even understood
how a parabolic mirror works as an image projector. Not until Brunelleschi's method,
probably via Alberti's writings, reached northern Europe after the 1430s did Flemish artists
(certainly Petrus Christus) employ optically correct perspective at all.

12. Meiss (2) I: 4; II: pl. 289.

7. Brunelleschi's Method

1. Krautheimer and Krautheimer-Hess I: 239 (see also his entire Chapter XVI: 229–53 on this subject); Summers (1): 514; (2): 64–65.

2. Saalman: 52 (*striscie di pergamene che si lievano per riquadrare le carte con numero d'abaco e caratte che Filippo intendeva per se medesimo*).

3. Vasari III: 142–143.

4. There is no record that Brunelleschi made scale drawings even in preparation for construction of the mammoth cupola over the Florentine Cathedral—at least none have survived. In truth, architectural drawings to scale were rarely employed in regular building practice anywhere in the Renaissance world until Vasari's own time in the late sixteenth century.

5. Edgerton (2): 93–114.

6. Ibid.: 143–52.

7. Trachtenberg.

8. See Meiss (1): 112–21, for a similar account of fourteenth-century St. Catherine of Siena who verified her miraculous visions because they appeared to her just as she had "seen them painted in churches."

9. Saalman: 44. Copyright 2008 by The Pennsylvania State University. Reproduced by permission of the publisher.

10. Alberti (3): 146. See Sinisgalli's note 235, p. 373, explaining how number six can be divided into perfect harmonic ratios: one being a sixth part of six, two a third, three a half, four two-thirds, and five five-sixths.

11. Gioseffi: 71–72; Klein.

12. Averlino I: 305–6.

13. Ibid. I: 304–5; II: 178v–179r. There are many sketches in the margins of the so-called *Medici Manuscript* here translated and published in facsimile by Spencer. This manuscript is not an autograph but was copied from the lost original by a professional scribe as a special dedication to Piero de' Medici probably around 1464. Unfortunately, the scribe, even though an expert calligrapher, was not adept at drawing, and, as Spencer admits, apparently failed to follow Filarete's verbal instructions as to what was to be represented. That is the case of the drawing supposedly to illustrate Filarete's instructions here, which clearly explain that a frontal view of a polygonal building in perspective also requires a centric point and two lateral projection points, one at either side along the same centric line, just as in my reconstruction of Brunelleschi's method in illustration 35.

8. Brunelleschi's Second Perspective Panel

1. Saalman: 44–45. Copyright 2008 by The Pennsylvania State University. Reproduced by permission of the publisher.

2. Antonine I, Tit. 17, cap. 1, cols. 831–34.

9. Brunelleschi's Heritage: Masaccio's *Trinity*

1. Vasari: 127, 143. Also, Edgerton (2): 27.

2. One *braccio* equals 58.36 centimeters; three equals 1.7508 meters.

3. Aiken: 95.

4. Masaccio at first apparently thought that painted figures as well as real-life figures when seen from below would appear to diminish in size, and so must be compensated for by depicting the head and upper body larger in proportion than the lower, as he did in his figure here of the Madonna—an idea, while unnecessary and incorrect in painting, that may have been suggested to him by Donatello who did achieve this effect successfully in three-dimensional sculpture.

5. Danti; see also Schlegel.

6. As measured from toe to the top of the cranium in its present position, the skeleton is only about five-feet-three-inches long (Aiken: 95). Although it cannot be verified under the current conditions by which this delicate fresco is being conserved by the Florentine superintendent of monuments, I believe that the skeleton, if it were repositioned along the edge of the tomb slab, would measure exactly three *braccia*, not from toe to the top of the cranium, however, but rather from toe to the cranial eye sockets.

7. See Danti, color plate VII, which shows—more clearly than my black and white reproduction in illustration 43—a dot at the center of the fictive alter top halfway between the two donor figures where the restorers found a small hole that they believe Masaccio used to place a nail from which he swung snap strings to establish the orthogonals of the vault. See also Polzer.

8. This has also been suggested by Ornella Casazza (80), a Florentine conservator who recently directed the cleaning and restoration of Masaccio's Brancacci Chapel frescoes.

9. Field: 43–61. Field found no measurable evidence in the fabric that Masaccio employed a gridded plan for constructing his perspective, but she also presumed that the picture plane itself began with the arched opening of the fictive chapel. The picture plane actually begins further forward at the edge of the projecting altar on which the donors kneel, at least a full virtual *braccio* in front of the recessed chapel entrance. Hence her estimated viewing distance of some twenty-two feet can't be right. Until another opportunity is afforded to make more measurements, especially of the bottom section of the fresco, I will hold that Masaccio began his construction with a gridded trapezoidal ground plan just as did Brunelleschi, and that his viewing distance, at least for foreshortening the squares of his ground level "floor" extending under the entire ensemble, was no more than the width of the painting itself (211.6 centimeters or nearly seven feet), in the same proportion as with Brunelleschi's Baptistery panel. Incidentally, Jane Aiken, for other reasons, has estimated exactly the same viewing distance (Aiken: 94–95).

10. See Gilbert for further analysis, especially in relation to other contemporaneous Florentine paintings. Antonino's original text in his *Summa*: III, Tit. 8, cap. iv, col. 322.

11. The Trinity was often represented in this latter form, and Antonino had probably seen examples in Florence. Such a peculiar rendition had its roots in pre-Christian worship, and pagan deities were sometimes so interpreted. Although many of these images may have existed in the archbishop's time, most were probably destroyed during the Counter-Reformation after Pope Urban VIII issued his specific bull condemning them as overtly heretical in 1628. For some peculiar reason, however, varied forms of three-headed Trinity representations persist to this day in the religious art of the Hispanic Americas.

12. Perrig.
13. Antonine I, Tit. III, cap. iii, col. 117.

10. "Oh, che dolce cosa è questa prospettiva!"

1. Vasari: 72.
2. Ibid.: 63.
3. Or what Michael Kubovy has called the "robustness" of perspective.
4. Bailey: 90–91.
5. Hartt: 177.
6. Shearman (2): 61.
7. Vasari: 127.
8. Shearman (1).
9. See Gardner von Teufel; and Strehlke: 89–109, who also criticize Shearman's reconstruction, but only because his hypothesized figural arrangement is so out of character with similar altarpieces of its time. While Gardner von Teufel noted Masaccio's low "vanishing point" between the Virgin's knees, she said nothing more about the way the artist employed perspective in arranging his composition. Neither scholar explained what caused the peculiar shadow markings to extend into the Madonna panel. In fact, both Gardner von Teufel and Strehlke concluded that the Madonna panel, even though slightly cut down, was framed separately from those of the standing saints at the sides, just as in the earlier San Giovenale altarpiece. See Rowlands: 58, who dismisses Gardner von Teufel's argument and reaffirms Shearman's reconstruction, but again with no recognition of its perspective problems. Furthermore, none of these recent authors saw any relationship between Masaccio's Pisa panel and Fra Angelico's San Domenico Altar, nor did they observe the singular clue hinting at that relationship, the unique elliptical halo behind the Christ child's head in each.
10. Summers (2).

11. More Masaccio, Masolino, and Even Fra Angelico

1. Alberti (3): 151–52.
2. A patch of *intonaco* just covers a nail hole in the wall beneath the frescoed head of Christ, which seems to have been the point from which Masaccio measured his major orthogonals. It has often been noticed that this head of Christ seems more in the style of Masolino than Masaccio himself. In fact, some have suggested that Masaccio possibly and purposely left this place uncovered with *intonaco* until the whole scene of the *Tribute Money* was painted, intending to paint in Jesus's face last of all. Either because he did not return or did not choose to do the face himself, it was left for Masolino to complete this critical iconic detail.
3. Kanter and Palladino: 3–12.
4. Exhibited between October 26, 2005, and January 29, 2006.
5. Kanter and Palladino: 72–73.
6. Ibid.: 73.

7. Ibid.: 59–64.

8. Fra Angelico did the same. See his *Madonna Enthroned* in Frankfurt; Kanter: 92.

9. See chapter 9, note 11.

10. For instance, his *Madonna di Cedri*, Museo Nazionale di San Matteo, Pisa, currently dated ca. 1415–17, but which I believe must have been done several years later. See Kanter: 11–12.

11. Fra Angelico with assistants apparently did emphasize the elliptical halo once more in several scenes on the Silver Chest, now in the Museo San Marco, ca. 1450. There is evidence, however, that the Chest was much repainted, and that the perspective haloes may have been added.

12. Kanter and Palladino: 84–87.

13. Ibid.: 74–75.

14. Alberti (3): 196.

15. Rubin (1), (2); Holmes; and Marcia Hall (personal communication). Fra Angelico's equivocation does not seem to have had anything to do with fear of idolatry, however. Although the danger of idol worship deeply concerned patrons of religious art in the earlier Middle Ages (see Kessler), the matter seems to have been resolved (at least casuistically) in the mind of Fra Antonino, who wrote in his *Summa* (II, Tit. 12,, cap. 1, col. 1136): "Images of saints are made in church, however, not to show [people] reverence of worship, but to impress their excellence effectively on the minds of men." In other words, viewers of religious art must not worship the image itself but concentrate only on the message of the holy narrative (*historia*).

16. Many thanks to my Williams College graduate students, Joshua O'Driscoll and Kori Lisa Yee Litt, who both wrote excellent papers on Fra Angelico's early paintings, sharing with me many of their own ideas on the painter's prescient use of perspective.

17. On the assumption, no longer acceptable, of Fra Angelico's naïveté, see Pope-Hennessy.

12. Alberti's Method

1. Ackerman: 87, note 1; see also Alberti (3): Introduction: 25–35.

2. There was also apparently an edition published in Nuremberg in 1511, but no copies are known to exist today, thus indicating that it had little circulation and no important influence. See Veltman (2): 6.

3. Alberti (3): 143–52.

4. Edgerton (1).

5. See Green and Green for a review of all the arguments among both art historians and mathematicians concerning Alberti's "distance point"; also Edgerton (1).

6. For a more detailed discussion of Pisanello's drawing, and how the artist modified his drawing, slightly but deliberately altering Alberti's directions to please his own aesthetic taste, see Edgerton (2): 52–54.

13. Alberti's Window

1. Alberti (3): 176–78.

2. Ibid.: 231.

3. Ibid.: 222.
4. Ibid.: 203.
5. Grafton (2): 30.
6. As translated by K. Clark: 184. The original Italian is in Alberti (1): cii–civ.

14. Alberti's Legacy

1. The *Stanza della Segnatura* suffers a Vasarian misnomer, since it was originally intended to be Pope Julius's library wherein to store (in low cupboards) his private collection of books.
2. Alberti (3): 227–31. See also Ackerman: 59–97.
3. Kaufmann.
4. Hall (1): 14–20, especially chapter 3.
5. Plato: lines 19b–22e.
6. See Edgerton (5).
7. The presence of this figure definitely shows how familiar Raphael was with Alberti's ideas. In *De Pictura*, Alberti advised "that in the *historia* there is someone who informs the spectators of things that unfold; . . . or invites you with his hand . . . to observe": Alberti (3): 212.
8. Redig de Campos: 10.
9. Serlio's popular treatise, originally published in Italian as *Sette libri dell'architettura* (Venice, 1537), was reprinted in numerous editions and in many European languages throughout the sixteenth and following centuries. Regarding Spanish resistance to perspective, see Edgerton (6): 121.
10. Ibid.: 121–27 for a much fuller explanation of these two engravings, especially in their sixteenth-century colonial Mexican context.
11. Ibid.: 107–71.

15. Galileo's "Perspective Tube"

1. A longer version of this chapter, with more complete footnotes and citations, is in Edgerton (5): 223–54.
2. Bredekamp.
3. Panofsky (2): 32–37.
4. Although the moon was rarely depicted objectively by Renaissance artists, a few instances do exist, the earliest perhaps by Jan Van Eyck (see Montgomery); the most "scientific" by Leonardo da Vinci (see Reaves).
5. Van Helden: 44.
6. Dupré: 369.
7. Galilei: 35–39.
8. Gingerich and Van Helden.
9. Ibid.
10. There is no way Galileo could have made such careful pen-and-wash studies during his exciting first moments at the telescope, as anyone who has ever stood in the cold, windswept tower of San Giorgio Maggiore (Galileo's open "observatory") should

quickly understand. Like any seventeenth-century "landscape painter," Galileo returned to the studio to finish his pictures, based on remembered impressions, verbal notes, and hasty diagrams. Plein air painting, after all, was not invented until the nineteenth century.

11. Galilei: 40.

12. See Reeves; Bredekamp.

13. Galilei: 41–43.

14. See Alexander, who argues that Hariot really did comprehend the rugged surface of the moon, but rather rendered what he saw only according to the contemporaneous graphic conventions of a mapmaker. Hariot indeed was trained as a cartographer and had accompanied Sir Walter Raleigh to the new American colony of Virginia in 1585 where he carefully and accurately sketched the coastal shoreline. Nevertheless, he still followed the style of the traditional medieval portolan chart, and thus, when he observed the moon, remained more conscious of its linear litoral-like features than of its mountains and valleys in chiaroscuro relief.

15. Regarding this fresco, see Van Helden and Booth; Ostrow (1), (2).

Postface: *Post perspectivam*

1. This possibility did not escape suspicious Protestants during the Reformation. John Donne the English poet, after hearing of Galileo's discoveries, regarded the whole matter as a Jesuit plot and so penned the following satirical tract, *Ignatius his Conclave*, in 1611: "I will write the Bishop of Rome: he shall call Galileo the Florentine who by this time hath thoroughly instructed himself of all the hills, woods, and cities in the moon. And now being grown to more perfection in his art, he shall have made new glasses, and with these having received a hallowing from the pope, he may draw the moon, floating like a boat upon the water, as near the Earth as he will. And thither (because they ever claim that those employments of discovery belong to them) shall the Jesuites be transferred, and easily unite and reconcile the Lunatique Church to the Roman Church. And without doubt, after the Jesuites have been there a little while, there will soon grow naturally a Hell in that world, over which you [?] Ignatius Loyola shall have dominion."

2. For more on the Italian (Sienese) engineers, Taccola and Francesco di Giorgio Martini, see Edgerton (5): 125–39.

3. See Friedberg for further intriguing thoughts on how Alberti's window still influences our visual media, from movies to Microsoft.

4. Lynch and Edgerton (1), (2).

5. Ibid.

Bibliography

Ackerman, James S. *Distance Points: Essays in Theory and Renaissance Art and Architecture*. Cambridge: MIT Press, 1994.

Aiken, Jane Andrews. "The Perspective Construction of Masaccio's Trinity Fresco and Medieval Astronomical Graphics." In *Masaccio's Trinity*, edited by Rona Goffen, 90–108. Cambridge: Cambridge University Press, 1998.

Alberti, Leon Battista (1). *Opere volgari de Leon Battista Alberti*, vol. 1. Edited by Anicio Bonucci. Florence: Tipografia Galileiana, 1843.

———(2). *Leon Battista Alberti On Painting*. Translated by Cecil Grayson, with introduction and notes by Martin Kemp. London: Penguin Books, 1991.

———(3). *Il nuovo De Pictura di Leon Battista Alberti—The New De Pictura of Leon Battista Alberti*. Edited and translated by Rocco Sinisgalli. Rome: Università di Roma La Sapienza, 2006.

Alexander, Amir. "Lunar Maps and Coastal Outlines: Thomas Hariot's Mapping of the Moon." *Studies in the History and Philosophy of Science* 29, no. 3 (1998): 345–68.

Alhacen: See Smith; Witelo.

Alighieri, Dante. *Il convivio*. In *Le opera de Dante Alighieri*, edited by Edward Moore and Paget Toynbee. Oxford: Clarendon Press, 1924.

Antonine, Saint. *Sancti Antonini Summa Theologica* (facsimile of 1740 Verona edition). 4 vols. Graz: Akademische Druck u. Verlagsanstalt, 1959.

Averlino, Antonio di Piero (Filarete). *Filarete's Treatise on Architecture*. 2 vols. Translated and edited by John R. Spencer. New Haven: Yale University Press, 1965.

Bacon, Roger (1). *The Opus Majus of Roger Bacon*. 2 vols. Edited by John H. Bridges. Oxford: Clarendon Press, 1897.

———(2). *De multiplcatione specierum*. In *The Opus Majus of Roger Bacon*, edited by John H. Bridges, vol. 2: 405–552. Oxford: Clarendon Press, 1897.

———(3). *The Opus Majus of Roger Bacon*, 2 vols. Translated and edited by Robert Belle Burke. Philadelphia: University of Pennsylvania Press, 1928.

Bailey, Gauvin Alexander. *Art on the Jesuit Missions in Asia and Latin America, 1542–1773*. Toronto: University of Toronto Press, 1999.

Baron, Hans. *The Crisis of the Early Italian Renaissance: Civic Humanism and Republican Liberty in an Age of Classicism and Tyranny*. 2 vols. Princeton: Princeton University Press, 1955.

Baxandall, Michael. *Painting and Experience in Fifteenth-Century Italy: A Primer in the Social History of Pictorial Style*. London: Oxford University Press, 1988.

Bialostocki, Jan. "Man and Mirror in Painting: Reality and Transience." In *Studies in Late Medieval and Renaissance Painting in Honor of Millard Meiss*, edited by Irving Lavin and John Plummer, vol. 1: 61–73. New York: New York University Press, 1977.

Biernoff, Suzannah. *Sight and Embodiment in the Middle Ages*. New York: Palgrave Macmillan, 2002.

Boraso, Stefano. *Brunelleschi 1420: Il paradigma prospettico di Filippo de Ser Brunellesco: il "caso" delle tavole sperimentale ottico-prospettiche*. Padua: Libreria Progetto, 1999.

Bredekamp, Horst. "Gazing Hands and Blind Spots: Galileo as Draftsman." *Science in Context* 13 (2000): 423–62.

Bryson, Norman. *Vision and Painting: The Logic of the Gaze*. New Haven: Yale University Press, 1983.

Casazza, Ornella. "Masaccio's Techniques and Problems of Conservation." In *Masaccio's Trinity*, edited by Rona Goffen, 65–90. Cambridge: Cambridge University Press, 1998.

Cennino Cennini. *The Craftsman's Handbook: Il libro dell' arte*. Translated and edited by Daniel V. Thompson Jr. New Haven: Yale University Press, 1933.

Clark, David L. "Optics for Preachers: The *De oculo morali* of Peter of Limoges." *Michigan Academician* 9 (1977): 329–43.

Clark, Kenneth. 1944. "Leon Battista Alberti on Painting." *Proceedings of the British Academy* 30 (1944): 1–20.

Crary. Jonathan. *Techniques of the Observer: On Vision and Modernity in the Nineteenth Century*. Cambridge: MIT Press, 1990.

Crombie, Alistair C. *Robert Grosseteste and the Origins of Experimental Science, 1100–1700*. Oxford: Clarendon Press, 1953.

Damisch, Hubert. 1995. *The Origin of Perspective*. Cambridge: MIT Press.

Danti, Cristina, ed. *La Trinità di Masaccio: Il restauro dell'anno Duemila*. Florence: Edifir, 2002.

Denery, Dallas G. *Seeing and Being Seen in the Later Medieval World: Optics, Theology, and Religious Life*. London: Cambridge University Press, 2005.

Dupré, Sven. "Galileo's Telescope and Celestial Light." *Journal of the History of Astronomy* 34 (2003): 369–99.

Edgerton, Samuel Y. (1). "Alberti's Perspective: A New Discovery and a New Evaluation." *Art Bulletin* 48 (1966): 367–78.

———(2). *The Renaissance Rediscovery of Linear Perspective*. New York: Basic Books, 1975; Harper Icon, 1976.

———(3). "Linear Perspective and the Western Mind: The Origins of Objective Representation in Art and Science." *Cultures* 3, no. 3 (1976): 77–194.

———(4). "Mensurare temporalia facit Geometria spiritualis: Some Fifteenth-Century Italian Notions about When and Where the Annunciation Happened." In *Studies in Late Medieval and Renaissance Painting in Honor of Millard Meiss*, edited by Irving Lavin and John Plummer, vol. 1: 115–130. New York: New York University Press, 1977.

——(5). *The Heritage of Giotto's Geometry: Art and Science on the Eve of the Scientific Revolution.* Ithaca: Cornell University Press, 1991.

——(6). *Theaters of Conversion: Religious Architecture and Indian Artisans in Colonial Mexico.* Albuquerque: University of New Mexico Press, 2001.

Elkins, James. *The Poetics of Perspective.* Ithaca: Cornell University Press, 1994.

Enciclopedia dantesca. Vols. 2, 4. Rome: *Enciclopedia Italiana,* nd.

Field, J. V. *The Invention of Infinity: Mathematics and Art in the Renaissance.* London: Oxford University Press, 1997.

Filarete: see Averlino.

Friedberg, Anne. *The Virtual Window: From Alberti to Microsoft.* Cambridge: MIT Press, 2006

Galilei, Galileo. *Sidereus nuncius or The Sidereal Messenger.* Translated by Albert Van Helden. Chicago: University of Chicago Press, 1989.

Gardner, Howard. *Artful Scribbles: The Significance of Children's Drawings.* New York: Basic Books, 1980.

Gardner von Teuful, Crista. "Masaccio and the Pisa Altarpiece: A New Approach." *Jahrbuch der Berliner Museen* 19 (1977): 27–68.

Ghiberti, Lorenzo (1). *I commentarii.* Edited by Ottavio Morisani. Naples: Riccardo Riccardi, 1947.

——(2). *Lorenzo Ghibertis Denkwürdigkeiten (I commentarii).* 2 vols. Translated by Julius von Schlosser. Berlin: Julius Bard Verlag, 1912.

Gibson, Eleanor J., and Anne D. Pick. *An Ecological Approach to Perceptual Learning and Development.* London: Oxford University Press, 2000.

Gibson, James J. *The Perception of the Visual World.* Boston: Houghton Mifflin, 1950.

Gilbert, Creighton. "The Archbishop on the Painters of Florence." *Art Bulletin* 41 (1959): 75–87.

Gingerich, Owen, and Albert Van Helden. "From Occhiale to Printed Page: The Making of Galileo's *Sidereus nuncius.*" *Journal of the History of Astronomy* 34 (2003): 251–67.

Gioseffi, Decio. *Perspectiva artificialis: Per la storia della prospettiva, spigolature e appunti.* Trieste: Istituto di Storia dell'Arte Antica e Moderna, Università degli Studi di Trieste, 1957.

Goldberg, Benjamin. *The Mirror and Man.* Charlottesville: University of Virginia Press, 1985.

Golomb, Claire. *The Child's Creation of a Pictorial World.* Mahwah, N.J.: Lawrence Erlbaum, 2004.

Gombrich, E. H. *Art and Illusion: A Study in the Psychology of Pictorial Representation.* Princeton: Princeton University Press, 1969.

Goodman, Nelson. *Languages and Art: An Approach to a Theory of Symbols.* Indianapolis: Bobbs-Merrill, 1968.

Grafton, Anthony (1). *Leon Battista Alberti, Master Builder of the Italian Renaissance.* New York: Hill and Wang, 2000.

——(2). *Bring Out Your Dead. The Past as Revelation.* Cambridge: Harvard University Press, 2001.

Grande Dizionario della Lingua Italiana. Vol. 14. Edited by Salvatore Battaglia. Turin: Unione Tipografico-Editrice, 1986.

Green, Judy, and Paul S. Green. "Alberti's Perspective: A Mathematical Comment." *Art Bulletin* 69, no. 4 (1987): 641–45.

Hall, Marcia B. (1). *After Raphael: Painting in Central Italy in the Sixteenth Century.*
Cambridge: Cambridge University Press, 1999.
———(2). *The Sacred Image in the Renaissance.* Forthcoming.
Harries, Karsten. *Infinity and Perspective.* Cambridge: MIT Press, 2001.
Hartt, Frederick. *History of Italian Renaissance Art: Painting, Sculpture, Architecture.* 4th
edition, revised by David G. Wilkins. Englewood Cliffs, N.J.: Prentice-Hall, 1994.
Henderson, Linda Dalrymple. *The Fourth Dimension and Non-Euclidian Geometry in
Modern Art.* Princeton: Princeton University Press, 1983.
Hockney, David (with Charles Falco). *Secret Knowledge: Rediscovering the Lost Techniques
of the Old Masters.* London: Viking Studio, 2001.
Holmes, Megan. *Fra Lippo Lippi: The Carmelite Painter.* New Haven: Yale University
Press, 1999.
Holton, Gerald. *Thematic Origins of Scientific Thought: Kepler to Einstein.* Cambridge:
Harvard University Press, 1988.
Ivins, William H., Jr. *Prints and Visual Communication.* Cambridge: MIT Press, 1953.
Kanter, Lawrence, and Pia Palladino, eds. *Fra Angelico.* Metropolitan Museum of Art
Series. New Haven: Yale University Press, 2005.
Kaufmann, Thomas DaCosta. *The Mastery of Nature: Aspects of Art, Science, and Human-
ism in the Renaissance.* Princeton: Princeton University Press, 1993.
Kellogg, Rhoda. *Analyzing Children's Art.* Palo Alto: National Press Books, 1970.
Kemp, Martin. *The Science of Art.* New Haven: Yale University Press, 1990.
Kessler, Herbert L. *Neither God nor Man: Words, Images, and the Medieval Anxiety about
Art.* Freiburg: Rombach Verlag, 2007.
Klein, Robert. "Études sur la perspective à la Renaissance: 1956–1963." *Bibliothèque
d'Humanisme et Renaissance: Travaux et Documents* 25, no. 3 (1963): 577–87.
Krautheimer, Richard, and Trude Krautheimer-Hess. *Lorenzo Ghiberti.* 2 vols. Princ-
eton: University of Princeton Press, 1970.
Kubovy, Michael. *The Psychology of Perspective and Renaissance Art.* Cambridge: Cam-
bridge University Press, 1985.
Kuhn, Jehane R. "Measured Appearances: Documentation and Design in Early
Perspective Drawing." *Journal of the Warburg and Courtauld Institutes* 53 (1990):
114–32.
Jex-Blake, K., trans. *The Elder Pliny's Chapters on the History of Art.* Commentary and
notes by E. Sellers. Chicago: Argonaut, 1968.
Lindberg, David C. (1). *A Catalogue of Medieval and Renaissance Optical Manuscripts.*
Toronto: Subsidia Mediaevalia, no. 4, 1975.
———(2). *Theories of Vision from Al-Kindi to Kepler.* Chicago: University of Chicago Press,
1976.
———(3). *Roger Bacon's Philosophy of Nature: A Critical Edition with English Translation,
Introduction and Notes of De multiplicatione specierum . . .* Oxford: Clarendon Press,
1983.
Lynch, Michael, and Samuel Y. Edgerton (1). "Aesthetics and Digital Image Processing:
Representational Craft in Contemporary Astronomy." In *Picturing Power: Visual
Depiction and Social Relations* (Sociological Review Monograph 35), edited by Gordon
Fyfe and John Law, 184–221, London: Routledge, 1988.
———(2). "Abstract Painting and Astronomical Image Processing." In *Aesthetics and
Science: The Elusive Synthesis,* edited by A. I. Tauber, 103–24. Dordrecht, Nether-
lands: Kluwer Academic Publishers, 1996.

Macfarlane, Alan, and Gerry Martin. *Glass: A World History.* Chicago: University of Chicago Press, 2002.

Massey, Lyle. "Configuring Spatial Ambiguity: Picturing the Distance Point from Alberti t' Anamophoses." In *The Treatise on Perspective, Published and Unpublished* (Studies in the History of Art 59), edited by Lyle Massey, 161–75. New Haven: Yale University Press, 2003.

Meiss, Millard (1). *Painting in Florence and Siena after the Black Death: The Arts, Religion, and Society in the Mid-Fourteenth Century.* New York: Harper and Row, 1951.

——(2). *French Painting in the Time of Jean de Berry: The Late XIV Century Patronage of the Duke.* 2 vols. London: Phaidon, 1967.

Melchior-Bonnet, Sabine. *The Mirror: A History.* Translated from the French by Katharine H. Jewett. New York: Routledge, 2001.

Mignolo, Walter D. *The Darker Side of the Renaissance: Literacy, Territoriality, and Colonization.* Ann Arbor: University of Michigan Press, 1995.

Montgomery, Scott L. "The First Naturalistic Drawings of the Moon: Jan Van Eyck and the Art of Observation." *Journal for the History of Astronomy* 25 (1994): 317–20.

New American Bible. Translated from the Original Languages with Critical Use of All the Ancient Sources. Washington, D.C.: World Catholic Press, 1970.

Novum Glossarium Mediae Latinitis ab Anno DCCC Usque ad Annum MCC (pea-pezzola). Geneva: Hafnia, 2000.

Ostrow, Steven F. (1). "Cigoli's *Immacolata* and Galileo's Moon: Astronomy and the Virgin in Early Seicento Rome." *Art Bulletin* 78 (1996): 218–35.

——(2). *Art and Spirituality in Counter-Reformation Rome: The Sistine and Pauline Chapels in S. Maria Maggiore.* Cambridge University Press, 1996.

Panofsky, Erwin (1). *Early Netherlandish Painting: Its Origins and Character.* 2 vols. Cambridge: Harvard University Press, 1958.

——(2). *Renaissance and Renascences in Western Art.* Stockholm: Almquist and Wiksell, 1960.

——(3). *Perspective as Symbolic Form.* Translated from the 1927 German edition by Christopher S. Wood. New York: Zone Books, 1993.

Parkhurst, Charles. "Giotto's Arena Chapel Frescoes and Religious Theater in His Time." *Mind's Eye* (Spring 2002): 38–55.

Parronchi, Alessandro. *Studi su la dolce prospettiva.* Milan: Aldo Martello, 1964.

Pecham, John. *John Pecham and the Science of Optics: Perspectiva communis.* Translated by David C. Lindberg. Madison: University of Wisconsin Press, 1970.

Pendergrast, Mark. *Mirror-Mirror: A History of the Human Love Affair with Reflection.* New York: Basic Books, 2003.

Perrig, Alexander. "Masaccios Trinita und der Sinn der Zentralperspektive." *Marburger Jahrbuch für Kunstgeschichte* 21 (1986): 11–45.

Pfeiffer, F., ed. *Meister Eckhart.* London: J. M. Watkins, 1931.

Plato. *Philebus.* Edited and translated by Robin Waterfield. Harmondsworth: Penguin, 1982.

Pliny (the Elder). *Natural History.* Vol. 9, books 33–35. Edited and translated by H. Rackham. Loeb Classical Library 394. Cambridge: Harvard University Press, 1952.

Polanyi, Michael. *Personal Knowledge: Towards a Post-Critical Philosophy.* London: Routledge and Kegan Paul, 1958.

Pollitt, J. J. *The Ancient View of Greek Art: Criticism, History, and Terminology.* New Haven: Yale University Press, 1974.

Polzer, Joseph. "The Anatomy of Masaccio's *Holy Trinity*." *Jahrbuch der Berliner Museen* 13 (1971): 18–59.

Pope-Hennessy, John. *Fra Angelico*. London: Phaidon, 1952.

Reaves, Gibson. "Leonardo da Vinci's Drawings of Surface Features of the Moon." *Bulletin of the American Astronomical Society* 17 (1985): 844–65.

Redig de Campos, Dioclecio. *The "Stanze" of Raphael*. Rome: Del Turco, 1968.

Reeves, Eileen. *Painting the Heavens: Art and Science in the Age of Galileo*. Princeton: Princeton University Press, 1997.

Romanshyn, Robert R. *Technology as Symptom and Dream*. London: Routledge, 1989.

Rowlands, Eliot W. *Saint Andrew and the Pisa Altarpiece*. Los Angeles: Getty Museum Studies on Art, 2003.

Rubin. Patricia (1). "Hierarchies of Vision: Fra Angelico's Coronation of the Virgin from San Domenico, Fiesole." *Oxford Art Journal* 27, no. 2 (2004): 137–51.

———(2). *Images and Identity in Fifteenth-Century Florence*. New Haven: Yale University Press, 2007.

Saalman, Howard, ed. *The Life of Brunelleschi by Antonio di Tuccio Manetti*. Translated from the Italian by Catharine Enggass. University Park: Pennsylvania State University Press, 1970.

Schechner, Sara. "Between Knowing and Doing: Mirrors and Their Imperfections in the Renaissance." *Early Science and Medicine* 10 (2005): 137–62.

Schlegel, Ursula. "Observations on Masaccio's Trinity Fresco in Santa Maria Novella." *Art Bulletin* 45, no. 1 (1963): 19–33.

Schuler, Stefan. *Vitruv im Mittelalter: Die Rezeption von "De architectura" von der Antike bis in die frühe Neuzeit*. Cologne: Böhlau, 1999.

Schwarz, Heinrich. "The Mirror of the Artist and the Mirror of the Devout." In *Studies in the History of Art Dedicated to William E. Suida on His Eightieth Birthday*, 90–105. London: Phaidon, 1959.

Sgrilli, Bernardo Sansone. *Descrizione e studi dell'insigne fabbrica di Santa Maria del Fiore . . . .* Florence, 1733.

Shearman, John (1). "Masaccio's Pisa Altarpiece: An Alternative Reconstruction." *Burlington Magazine* 108 (1966): 449–55.

———(2). *Only Connect: Art and the Spectator in the Italian Renaissance*. Princeton: Princeton University Press, 1992.

Silver, Matthew. *A Tale of Two Paradigms: The Congruence of Art and Science in Renaissance Thought and Modern Perception*. Senior thesis, Williams College, Williamstown, Mass., 2001.

Smith, Mark A. *Alhacen's Theory of Visual Perception: A Critical Edition, with English Translation and Commentary of the First Three Books of Alhacen's 'De Aspectibus,' the Medieval Latin Version of Ibn al-Haytham's 'Kitāb al Manāzir'.* 2 vols. Philadelphia: American Philosophical Society, 2001.

Steinberg, Leo, and Samuel Y. Edgerton. "How Shall This Be?" *Artibus et Historiae* 16, VIII (1987): 45–53.

Strehlke, Carl B. *The Panel Paintings of Masaccio and Masolino: The Role of Technique*. Milan: 5 Continents Editions, 2002.

Summers, David (1). *Real Spaces: World Art History and the Rise of Western Modernism*. London: Phaidon, 2003.

———(2). *Vision, Reflection, and Desire in Western Painting*. Chapel Hill: University of North Carolina Press, 2007.

Tachau, Katherine H. "Seeing as Action and Passion in the Thirteenth and Fourteenth Centuries." In *The Mind's Eye: Art and Theological Arguments in the Middle Ages*, edited by Jeffery F. Hamburger and Anne-Marie Bouché. Princeton: Princeton University Press, 2006.

Tanturli, Giuliano. "Rapporti del Brunelleschi con gli ambienti letterari fiorentini." In *Filippo Brunelleschi: La sua opera e il suo tempo*, vol. 1: 125–45. Florence: Centro di, 1977.

Thesaurus Linguae Latinae. Vol. X.1, fasc. xi; vol. x.2, fasc. xi 4. Leipzig: H. B. Teubner, 1998–2004.

Thornton, Peter. *The Italian Renaissance Interior, 1400–1600*. New York: Harry N. Abrams, 1991.

Trachtenberg, Marvin. *Dominion of the Eye: Urbanism, Art, and Power in Early Modern Florence*. London: Cambridge University Press, 1997.

Van Helden, Albert. "The Telescope in the Seventeenth Century." *Isis* 65 (1974): 38–59.

Van Helden, Albert, and Sara E. Booth. "The Virgin and the Telescope: The Moons of Cigoli and Galileo." In *Galileo in Context*, edited by Jürgen Renn, 193–216. Cambridge, UK: Cambridge University Press, 2001.

Vasari, Giorgio. *Le vite de'più eccellenti pittori scultori e architettori*. Edited by Rosanna Bettarini and Paola Barocchi. Vol. 3. Florence: Sansoni Editore, 1971.

Veltman, Kim H. (1). *Abstract: The Sources of Perspective*. 1989. http://www.sumscorp.com/books/pdf/2004%20Sources%20of%20 Perspective.pdf.

———(2). *Electronic Media in the Study of Alberti*. 1995. www.mmi.unimaas.nl/people/Veltman/veltmanarticles/1988%20Perspective%20Courses.pdf.

Vignola, Giacoma Barozzi. *Le dve regole della prospettiva pratica . . . Con i comentarij del r. p. m. Egnatio Danti [. . .]*. Roma: Francesco Zannetti, 1583. Available at the Clark Art Institute Library.

Villani, Giovanni. *Cronache di Giovanni, Matteo, e Fillipo Villani*. Trieste, 1857.

Vulgata: Biblia sacra iuxta vulgatum versionem. 2 vols. Stuttgart: Würtembergische Bibelanstalt, 1969.

Weil, Mark. "The Devotion of the Forty Hours and Roman Baroque Illusion." *Journal of the Warburg and Courtauld Institutes* 37 (1974): 218–49.

Wesselofsky, A., ed. *Il Paradiso degli Alberti, Retrovi e ragionamenti del 1389 Romanza di Giovanni da Prato*. Bologna, 1867.

White, John. *The Birth and Rebirth of Pictorial Space*. 2nd edition. London: Faber and Faber, 1965.

Winner, Ellen. "Where Pelicans Kiss Seals." *Psychology Today* 20, no. 7 (1986): 24–27.

Witelo. *Perspectiva*. In *Opticae thesaurus Alhazeni Arabis libri septem . . . item Vitellonis Thuringopoloni libri X*. Edited by Friedrich Risner. Basel, 1572.

Woodward, David, and J. B. Harley, eds. (1). *The History of Cartography: Cartography in Prehistoric, Ancient, and Medieval Europe, and the Mediterranean*. 2 vols. Chicago: University of Chicago Press, 1987.

———(2). *The History of Cartography: Cartography in the Traditional and Islamic and South Asian Societies*. 2 vols. Chicago: University of Chicago Press, 1992–95.

Index

Abū 'Alī Al-Hasan Ibn Al-Haytham. *See* Alhacen
Accademia del Disegno ("Academy of Drawing"), 152–153, 154, 162
Adam and Eve, 13, 44
Aeschylus, 15
Agatharcus of Athens, 15
Aiken, Jane, 80, 180n6, 180n9
Albategnius, crater on the moon, 162, 167
Alberti, Leon Battista, 7, 20, 22, 30, 40, 61, 67, 75, 80, 100, 106, 114, 133, 136, 141, 147, 151, 183chap14n7; his *De Pictura*: dedicated to Brunelleschi, 117; English translation by Rocco Sinisgalli, 175n6; Alberti's perspective method a codification of Brunelleschi's principles, 8; Alberti's perspective proof, 121–122; adumbrated by Masaccio, 106; Alberti's preference for perspective serving a moral purpose rather than religious, 7–8, 129; his views on the use of chiaroscuro, 9, 136; his veil, 126–127; his advice to artists about antique subject matter: depicting dead Meleager, 114, 131, Inachus's daughter and Alexander's horse, Bucephalus, 129, 135; Alberti's definition of horizon line isocephaly, 100, 122–123; his advice against depicting too many figures, 130
Alexander VI, Pope, 133
Alhacen, 21–22
anamorphoses, 71
angels: surrounding the Madonna in Masaccio's paintings, 93–98; compared to Fra Angelico's angels, 113; Angel Gabriel, 32, 34

angle of incidence, angle of reflection. *See* mirrors
antiquarianism in Rome, 135
Antonino Pierozzi. *See* Fra Antonino
Apelles, 40
Apocalypse, 13
Arab. *See* Islam
Arch of Titus, Rome, 135–136
Arena Chapel, Padua, *Life of the Virgin* fresco cycle by Giotto, 17–19
Aristotle, 139, 156
Ark of the Covenant, 79
Armstrong, Neil, 168
Arte di Calimala "Wool Merchants' Guild," 48, 49
Artes Liberales, 42–43
Assisi. *See* San Francesco Basilica
atmospheric perspective, 91
Averlino, Antonio. *See* Filarete
axis perpendicularis, 25, 31, 48, 50

Bacon, Francis, 156
Bacon, Roger: on the need for more geometrical application to reinvigorate Christianity, 15; his reference to "Latins" as experts in geometry, 15–16; adumbration (inadvertent) of later geometric perspective in painting, 19, 21; his *species* theory explained in *De multiplicatione specierum*, 28; his treatise on *Perspectiva*, 30, 31; his *species* theory as applied in Lippi's *Annunciation*, 36–38
bandiera; "flag" of the San Giovanni *quartiere*, 46–47

mirrors, 6–7, 24–25, 34, 43; 50–55; angle of incidence, angle of reflection, 25; burning mirrors, 15; medieval manufacture of convex glass and flat silvered metal mirrors, 27; optical theory of, 22–25; believed to reflect and retain the power of holy relics, 24; *Speculum* as popular book title, 27; as artist's tool for correcting drawing defects, 127; Jesus as "mirror of nature," 148

Mohammed. *See* Islam

moon, 154–168

Murillo, Bartolomé Estabán, 156–157. *See also* Saint Mary, Virgin

Muses, 43

naïve: referring to art created without knowledge of linear perspective, 4, 72; as in children's art, 3; also in the arts of non-Western cultures, 4–5; as well as pre-Renaissance Western art, 5, 145

Narcissus, 24

Nardo di Cione, *Trinity*, 79

National Gallery of Art, London, 36, 93

Natura "Nature," 7, 8, 75, 112, 127, 129, 132, 145, 172

Nelli, Ottaviano, 62; his painting *Circumcision of Baby Jesus* revealing bifocal construction, 63

Newton, Sir Isaac, 167

oblique perspective, 58, 71, 72

one-point perspective. *See* frontal perspective

optics, 2, 13, 15, 21, 22, 55

Orsanmichele Church, Florence, 11, 116

orthogonal, 83, 124

orthographic projection, 54–55, 76, 90

Padua, 154, 159

Palazzo Signoria. *See* Palazzo Vecchio

Palazzo Vecchio, 5, 11, 50, 58, 69–70, 73, 74

Panofsky, Erwin, 5, 152

Paradiso, 44, 50

Parkhurst, Charles, 17

Pauline Chapel, Santa Maria Maggiore, Rome, 167

pavimentum "pavement," 131, 137

Pecham, John, 21, 31

Pelacani, Biagio. *See* Blasius of Parma

perspectiva artificialis; a medieval solution to a medieval problem, 20, 22, 30, 43, 145

perspectiva communis. See perspectiva naturalis

perspectiva naturalis, 22, 30, 41, 43, 47, 55, 60, 76, 98, 176chap3n3. *See also* visual pyramid

perspective glasse, 165

perspective tube, 154, 158

perspicere, perspectus, 48, 176n3chap3

perspicillum, 159

Peter of Limoges, his treatise *De oculo morali*, 31, 32

Philosophia, 42–43

Piero della Francesca, 41

Piero de'Medici. *See* Medici

Pierozzi. *See* Fra Antonino

Pintoricchio, Bernardino, 133

Pisa, 151, 153

Pisanello, 124–125; his perspective may reveal Masaccio's method for laying out the *Trinity*, 125

pixel. *See* digital image processing

Plato, 7, 72, 139–140

Pliny, 40, 52

Polanyi, Michael, 178n10

pontifex maximus, 134

prespettivo, 39, 109. *See also perspective naturalis*

prospectiva, 42–43. *See also perspective naturalis*

prospettiva. See perspectiva naturalis

Protogines, 40

"Pseudo Jacopino," his painting, *Vision of Saint Romuald*, 145, 146

Ptolemeus, Claudius "Ptolemy," 56; his *Geographia*, 56–57

quadrivium, 15

quartiere of San Giovanni, 46, 74

Raphael; most Albertian of all Renaissance artists, 133; his application of Albertian perspective, 137–138; his understanding of chiaroscuro, 139–141; his frescoes in the *Stanza della Segnatura: Theology*, sometimes called the *Disputà*, 137–139, 140–142; *Philosophy*, sometimes called the *School of Athens*, 137–140; his inclusion of a beckoning Albertian interlocutor, 141, 183n7; his Euclidian representation of heavenly space, 142

refraction of light rays, 22, 25

"relief-like" style, 136, 139–140, 149; as applied by Christian missionaries to make holy images appear "divinely present," 149–150

Renaissance Rediscovery of Linear Perspective, xiii, 56

Roman School of painting during the early Middle Ages, 15–16

Rome, at time of Raphael's arrival, 134–135

Rubin, Patricia, 116

Vignola, Giacomo Barozzi da, xiii
Virgo galaxy as depicted by electronic CCD, 172
visual axis. *See axis perpendicularis*
visual cone. *See* visual pyramid
visual pyramid, 22–23, 33, 47
Vita anonima, 132

Vitruvius: on chiaroscuro and "perspective" projection in theatrical stage settings, 16

water pump; drawings, 168–171
Wells, H. G., 168
White, John, 90
Witelo, 21

www.ingramcontent.com/pod-product-compliance
Lightning Source LLC
Chambersburg PA
CBHW020903180526
45163CB00007B/2607